IEE Power and Energy Series 41

Series Editors: Professor A. T. Johns
D. F. Warne

Electrical operation of electrostatic precipitators

Other volumes in this series:

Volume 1   **Power circuits breaker theory and design** C. H. Flurscheim (Editor)
Volume 2   **Electric Fuses** A. Wright and P. G. Newbery
Volume 3   **Z-transform electromagnetic transient analysis in high-voltage networks** W. Derek Humpage
Volume 4   **Industrial microwave heating** A. C. Metaxas and R. J. Meredith
Volume 5   **Power system economics** T. W. Berrie
Volume 6   **High voltage direct current transmission** J. Arrillaga
Volume 7   **Insulators for high voltages** J. S. T. Looms
Volume 8   **Variable frequency AC motor drive systems** D. Finney
Volume 9   **Electricity distribution network design** E. Lakervi and E. J. Holmes
Volume 10   **SF$_6$ switchgear** H. M. Ryan and G. R. Jones
Volume 11   **Conduction and induction heating** E. J. Davies
Volume 12   **Overvoltage protection of low-voltage systems** P. Hasse
Volume 13   **Statistical techniques for high-voltage engineering** W. Hauschild and W. Mosch
Volume 14   **Uninterruptible power supplies** J. D. St. Aubyn and J. Platts (Editors)
Volume 15   **Digital protection for power systems** A. T. Johns and S. K. Salman
Volume 16   **Electricity economics and planning** T. W. Berrie
Volume 17   **High voltage engineering and testing** H. M. Ryan (Editor)
Volume 18   **Vacuum switchgear** A. Greenwood
Volume 19   **Electrical safety: a guide to the causes and prevention of electrical hazards** J. Maxwell Adams
Volume 20   **Electric fuses, 2nd Edn.** A. Wright and P. G. Newbery
Volume 21   **Electricity distribution network design, 2nd Edn.** E. Lakervi and E. J. Holmes
Volume 22   **Artificial intelligence techniques in power systems** K. Warwick, A. Ekwue and R. Aggarwal (Editors)
Volume 23   **Financial and economic evaluation of projects in the electricity supply industry** H. Khatib
Volume 24   **Power system commissioning and maintenance practice** K. Harker
Volume 25   **Engineers' handbook of industrial microwave heating** R. J. Meredith
Volume 26   **Small electric motors** H. Moczala
Volume 27   **AC-DC power system analysis** J. Arrillaga and B. C. Smith
Volume 28   **Protection of electricity distribution networks** J. Gers and E. J. Holmes
Volume 29   **High voltage direct current transmission** J. Arrillaga
Volume 30   **Flexible AC transmission systems (FACTS)** Y.-H. Song and A. T. Johns (Editors)
Volume 31   **Embedded generation** N. Jenkins, R. Allan, P. Crossley, D. Kirschen and G. Strbac
Volume 32   **High voltage engineering and testing, 2nd Edn.** H. M. Ryan (Editor)
Volume 33   **Overvoltage protection of low voltage systems** P. Hasse
Volume 34   **The lightning flash** G. V. Cooray
Volume 35   **Contol Techniques drives and controls handbook** W. Drury (Editor)
Volume 36   **Voltage quality in electrical systems** J. Schlabbach, D. Blume and T. Stephanblome
Volume 37   **Electrical steels** P. Beckley
Volume 38   **The electric car: development and future of battery, hybrid and fuel-cell cars** M. H. Westbrook
Volume 39   **Power systems electromagnetic transients simulation** N. Watson and J. Arrillaga
Volume 40   **Advances in high voltage engineering** M. Haddad and D. Warne (Editors)

Electrical operation of electrostatic precipitators

Ken Parker

The Institution of Electrical Engineers

Published by: The Institution of Electrical Engineers, London, United Kingdom

The Institution of Electrical Engineers,
Michael Faraday House,
Six Hills Way, Stevenage,
Herts. SG1 2AY, United Kingdom
www.iee.org

British Library Cataloguing in Publication Data

Parker, K. R.
Electrical operation of electrostatic precipitators. – (IEE power & energy series; no. 41)
1. Electronic precipitators
I. Title II. Institution of Electrical Engineers
537.2

ISBN 0 85296 137 5

Typeset in the UK by RefineCatch Ltd, Bungay, Suffolk
Printed in the UK by MPG Books Limited, Bodmin, Cornwall

Contents

Preface xi
Acknowledgements xiii

1 The range and application of electrostatic precipitators 1
  1.1 Introduction 1
  1.2 Arrangement and basic operation of a precipitator 4
  1.3 Versatility 10
  1.4 Characteristics of the gases as they affect precipitator design 11
    1.4.1 Composition 11
    1.4.2 Temperature 12
    1.4.3 Pressure 13
    1.4.4 Gas flow rate 13
    1.4.5 Viscosity and density 14
  1.5 Characteristics of the suspended material and their possible impact on precipitator performance 14
    1.5.1 Concentration 14
    1.5.2 Composition and electrical resistivity 15
    1.5.3 Particle sizing 16
    1.5.4 Particle shape 17
    1.5.5 Particulate surface properties 18
  1.6 Sizing of electrostatic precipitators 19
  1.7 References 19

2 Fundamental operation of an electrostatic precipitator 21
  2.1 Introduction 21
  2.2 Ion production 22
  2.3 Particle charging 27
  2.4 Particle migration 29
  2.5 Particle deposition and removal from the collector electrodes 32
  2.6 Precipitator efficiency 33
  2.7 Practical approach to industrial precipitator sizing 34
  2.8 References 36

3   Factors impinging on design and performance   39
   3.1   Effect of gas composition   39
   3.2   Impact of gas temperature   40
   3.3   Influence of gas pressure   43
   3.4   Gas viscosity and density   44
   3.5   Impact of gas flow rate and gas velocity   44
   3.6   Gas turbulence   46
   3.7   The importance of gas distribution   48
         3.7.1   Correction by model testing   50
         3.7.2   Computational fluid dynamic approach   51
   3.8   Effect of particle agglomeration   54
   3.9   Particulate cohesivity effects   58
   3.10 References   61

4   Mechanical features impacting on electrical operation   63
   4.1   Production of ions   63
         4.1.1   Quasi-empirical relationships   66
   4.2   Discharge electrode forms   68
   4.3   Spacing of discharge electrodes   71
   4.4   Collector electrodes   72
   4.5   Specific power usage   73
   4.6   Precipitator sectionalisation   75
   4.7   High tension insulators   76
   4.8   Electrical clearances   78
   4.9   Deposit removal from the collector and discharge systems   80
         4.9.1   Impact of collector deposits on electrical operation 80
         4.9.2   Measurement of rapping intensity on dry precipitator
                 collectors   84
   4.10 Hopper dust removal   85
   4.11 Re-entrainment from hoppers   86
   4.12 References   86

5   Development of electrical energisation equipment   89
   5.1   Early d.c. energisation techniques   89
   5.2   Development of a.c. mains frequency rectification equipment   91
         5.2.1   Valve rectifiers   91
         5.2.2   Mechanical switch rectifier   92
         5.2.3   Metal oxide rectifiers   97
         5.2.4   Silicon rectifiers   98
   5.3   HT transformers   101
         5.3.1   Transformer losses   103
   5.4   Cooling of transformer and rectifier equipment   105
   5.5   Primary input control systems   107
         5.5.1   Manual methods   107
         5.5.2   Motorised methods   108

      5.5.3   Saturable reactors   109
      5.5.4   Silicon controlled rectifiers (SCRs)   110
  5.6 Automatic control systems   113
      5.6.1   Early current control   113
      5.6.2   Voltage control method   114
      5.6.3   Computer control methods   115
  5.7 Summary of developments   116
  5.8 References   117

6 Modern mains frequency energisation and control   119
  6.1 Basic operation of mains frequency equipment   119
  6.2 High voltage equipment supply ratings   123
      6.2.1   Mean precipitator current   124
      6.2.2   Primary r.m.s. current   125
      6.2.3   Precipitator peak voltage under no-load conditions   125
      6.2.4   Apparent input power   125
      6.2.5   Practical example   126
  6.3 Influence of the linear inductor   126
      6.3.1   The main function of the inductor   126
      6.3.2   Physical implementation of the linear inductor   128
  6.4 Automatic voltage control and instrumentation   129
      6.4.1   Secondary metering approach   130
      6.4.2   Primary metering system   131
      6.4.3   Opacity signal and full energy management   132
  6.5 Basic AVC control principles   135
  6.6 Back-corona detection and corona power control   140
  6.7 Specific power control   142
      6.7.1   Control by the use of intermittent energisation   142
      6.7.2   Supervisory computer control using a gateway approach   143
  6.8 Advanced computer control functions   144
  6.9 References   146

7 Alternative mains frequency energisation systems   147
  7.1 Two-stage precipitation   147
      7.1.1   Air cleaning applications using positive energisation   147
      7.1.2   Two-stage precipitation as applied to power plant
             precipitators   149
  7.2 Intermittent energisation   151
      7.2.1   Basic principles of intermittent energisation   152
      7.2.2   Comparison of IE with traditional d.c. energisation   153
      7.2.3   Collection efficiency evaluation   156
  7.3 Pulse energisation   159
      7.3.1   Introduction   159
      7.3.2   Electrical configurations   160
      7.3.3   Electrical operation with pulse chargers   164

7.4   Major operational aspects of pulse energisation   165
    7.4.1   Current control capabilities   166
    7.4.2   Current distribution on collectors   167
    7.4.3   Electrical field strength in the inter-electrode area   167
    7.4.4   Particle charging   168
    7.4.5   Power consumption   168
    7.4.6   Worked example of energy recovery   169
7.5   Collection efficiency   169
7.6   Typical applications using pulse energisation   170
7.7   References   171

8   High frequency energisation systems   175
8.1   Introduction   175
8.2   Development of switch mode power supply systems   176
    8.2.1   Expected operational improvements   176
    8.2.2   System requirements   177
    8.2.3   Design considerations   178
    8.2.4   SMPS circuit configurations   179
8.3   Input rectification stage   179
    8.3.1   Three phase six switch mode UPF converter   180
    8.3.2   Three phase boost type UPF rectifier   181
    8.3.3   Three phase full wave rectifier with a.c. side filtering   182
    8.3.4   Comments on the various input stage topologies   183
8.4   High frequency inverter stage   184
    8.4.1   PWM controlled 'H' bridge inverter   185
    8.4.2   Resonant converter   185
    8.4.3   Matrix converter   187
    8.4.4   Comments on the various inverter topologies   188
8.5   High voltage, high frequency transformer   188
    8.5.1   Parasitic capacitance   188
    8.5.2   Magnetic leakage flux   188
    8.5.3   Insulation and electric stress management within the
        transformer   189
    8.5.4   Corona effects   190
8.6   Transformer output rectification   190
8.7   Short duration pulse operation   190
8.8   Advantages of the SMPS approach to precipitator energisation   191
8.9   Review of the various topologies leading to a prototype
    SMPS development under a UK EPSRC grant   192
8.10 Operational field experience with SMPS precipitator energisation   192
    8.10.1 SMPS system of Supplier No. 1   194
    8.10.2 SMPS system of Supplier No. 2   196
    8.10.3 SMPS system of Supplier No. 3   204
    8.10.4 SMPS system of Supplier No. 4   209
    8.10.5 Advantages/conclusions reached from these field trials   212

8.11 References   213

**9 The impact of electrical resistivity on precipitator performance
and operating conditions**   215
  9.1 Particle composition   215
  9.2 Particle resistivity   216
  9.3 Measurement of particulate resistivity   217
  9.4 Resistivity effects on low temperature power station
      precipitators   220
  9.5 Correlation between precipitator performance and
      particulate resistivity   222
  9.6 Low resistivity ash and its effect on performance   226
  9.7 Equipment for flue gas conditioning   228
    9.7.1 Sulphur trioxide conditioning   228
    9.7.2 Ammonia conditioning   230
  9.8 Humidity conditioning   232
  9.9 Reducing inlet gas temperature   233
  9.10 Summary of resistivity effects on precipitator performance 233
  9.11 References   236

10 'On-line' monitoring, fault finding and identification   239
  10.1 Corrosion condition monitoring   239
  10.2 Electrical operating conditions   248
    10.2.1 TR control cubicle information   249
    10.2.2 TR set panel meter readings   249
    10.2.3 TR equipment lamp test   249
  10.3 Deposit removal from the internals   253
  10.4 DE and collector system voltage/current relationships   253
    10.4.1 Clean air load characteristic   254
    10.4.2 Operational curves with gas passing through the
        system   255
    10.4.3 Dirty air load test, without gas passing through system 258
  10.5 Collector/discharge electrode alignment   259
  10.6 High tension insulators   259
  10.7 Hoppers   260
  10.8 Gas distribution and air inleakage   261
  10.9 Changing inlet gas conditions   261
    10.9.1 Particle resistivity   261
    10.9.2 Particle sizing and opacity monitoring   262
  10.10 Systematic fault finding procedure   264
  10.11 References   264

Index   265

Preface

The electrostatic precipitator, in spite of its large footprint, remains one of the most cost effective means of controlling the particulate emission from large industrial process plant. International Environmental Agencies are currently requiring emissions of <50 mg Nm^{-3} for general applications and 10 mg Nm^{-3} where discharges are liable to prove injurious to health, e.g. heavy metal and potential carcinogenic materials. The current major application is to minimise atmospheric pollution and for some power generating plants, particulate removal efficiencies in excess of 99.8 per cent are not uncommon in order to meet legislative emissions.

Although electrostatic precipitators have been used for almost a century, the most significant changes, in terms of design and plant reliability, have been mainly concerned with the method of energising the electrode system. Following a description of the process physics and the large number of gas and particle characteristics that impact on the design and operation of a precipitator, the essential requirements of the electrical energisation system are identified, such that the precipitator installation can operate at maximum performance under optimum electrical conditions. Theoretically it will be shown that the maximum particulate collection efficiency is proportional to the applied voltage squared, that is, the precipitator needs to operate as close to the breakdown potential as possible. Unlike many electrical systems, those used for precipitator duties have normally to operate under short circuit conditions, which places significant electrical stress on all components, a situation that needs careful consideration in the basic design of the energisation equipment, particularly the high voltage transformer.

A review of the development of precipitator energisation is included, commencing with the early attempts using electrostatic voltage generators, then, through the intermediate rectification steps of mechanical, valve and metal oxide rectifiers and the use of tapped inlet transformers, motorised autotransformers and saturable reactors for primary voltage control, through to the present silicon rectifier and thyristor systems in current usage, along with the almost parallel development of automatic voltage controllers. These AVC systems were initially based on analogue primary current, secondary current and secondary voltage inputs, then, as A/D converters became available, the development of full digital controllers.

Possibly the most significant physical property of the particulate, as far as precipitator performance/efficiency is concerned, is its electrical resistivity. For particulates having resistivities in excess of $1e^{12}$ Ωcm they cannot readily lose their charge upon reaching the collector and produce a condition termed reverse or back ionisation. In this case, a voltage builds up on the collector element and gives rise to a positive ion discharge, which in moving towards the discharge electrode neutralises the negative charge carriers and modifies the electric field such as to significantly impact on particle migration and hence overall plant performance. Although the particulate resistivity can be reduced by flue gas conditioning approaches, the impact of reverse ionisation can be partially mitigated by using a discontinuous mode of energisation. That is, the application of either an intermittent or pulse charging method of energisation, where time is allowed for the collected charge to dissipate between successive energisation pulses. The basic design and operation of both intermittent and pulse charging forms of energisation are reviewed together with some typical results.

In the past decade, with further advances in silicon high speed switch technologies, a new form of energisation has become possible based on high frequency, high power switched mode forms of equipment. The present status of this switch mode power supply (SMPS) equipment is examined together with a review of its impact on precipitator energisation and design, since the SMPS approach produces an almost pure d.c. waveform, enabling higher operating voltages and hence performance levels to be achieved.

Although the mechanical design and electrical operation of present day precipitators leads to plant availabilities well in excess of 97 per cent, there are sometimes operational difficulties that adversely impact on performance, which fall under the general headings of corrosion, dust build up and mechanical damage to the precipitator internals and/or electrical equipment. The final chapter includes a breakdown of how some of these shortcomings can be readily identified from the installed instrumentation on the plant, thereby reducing the need to internally inspect the plant for the cause, but not necessarily its correction. With the advancement of automatic control computer programming it is envisaged that many of these operational difficulties will be automatically recognised and flagged to the operators, which should further improve plant availability.

In conclusion, the design of the energisation equipment has closely tracked the latest developments in electrical and control engineering practices to produce a highly efficient, reliable, low maintenance and largely hands off system, together with complex microprocessor programming leading to the complete automatic control of the whole plant.

K. R. Parker, *Consultant Engineer, Sutton Coldfield, UK*

Acknowledgements

Although some 40 years of personal experience in the field of electrostatic precipitation has been translated into the book, the author wishes to express his thanks and appreciation to Douglas Warne for his initial support and encouragement for the work and to Victor Reyes and Leif Lind for help with the computer and CFD sections. Thanks must also go to Paul Lefley of Leicester University, Martin Kirsten of Alstom Power, Victor Reyes of FLS Miljö s/a, Oywind Wetteland of Applied Plasma Physics and Helmut Herder of NWL Transformers for assistance with the switch mode power supply status chapter.

Thanks must also be acknowledged to Lodge Sturtevant, Leicester University, Alstom Power, FLS Miljö s/a, Applied Plasma Physics NWL Transformers and Corrosion Management for providing photographs of plant and equipment, all of which are suitably referenced.

The author would also like to acknowledge the kind permission of Kluwer Academic Publishers to reuse diagrams, figures and equations, which appeared in 'Applied Electrostatic Precipitation', ISBN 0 7514 02664, and to the International Electrostatic Precipitation Society to include diagrams which have appeared in their Conference Proceedings.

Thanks also must be accorded to those who have been involved in correcting the final manuscript, Ken Darby, who was my mentor and a source of inspiration for some 35 years, and my wife, Maureen, for her encouragement and support in the project.

Chapter 1

The range and application of electrostatic precipitators

1.1 Introduction

There has, in recent years, been an increasing awareness and world-wide recognition of the problems associated with atmospheric pollution; consequently, most industrialised nations have enacted legislation to limit uncontrolled emissions from all sources. This legislation is continuously reviewed and is becoming more stringent, such that particulate emissions, in particular, are presently controlled to a maximum of 50 mg Nm^{-3} for inert materials and 10 mg Nm^{-3} for substances considered hazardous to health and it is suspected that within the next decade these values will be further reduced.

One of the most suitable forms of arrestment device for meeting these levels of emission, particularly for process plants producing large gas flow rates, is the electrostatic precipitator. Following the installation of the first pulverised coal steam raising plant in the 1920s, where some 90 per cent of the ash can be carried forward with the waste gases, the electrostatic precipitator has been almost universally used to control particulate emissions in the power generation industry. To meet current emission regulations, fly ash collection efficiencies in excess of 99.8 per cent can be required when handling gas flows of up to 1000 $am^3 s^{-1}$.

The features of the electrostatic precipitation process, which produces an ideal vehicle for the removal of particulates, are as follows.

(1) Versatility – effective performance on a wide range of industrial processes. Can be designed to meet any required efficiency and sized for any gas flow rate.
(2) Designs can be produced to cover a temperature range from ambient up to 850 °C.
(3) Can collect particles over the complete size range spectrum.
(4) Dust is usually recovered in its original state. But plants can be designed to

operate as a wet phase device if required, particularly for gases close to, or at, dew point temperature.

(5) Low pressure loss – typically less that 1 mbar.
(6) Acceptable electrical power consumption for required efficiency level.
(7) Robust and reliable construction – life expectancy >20 years.
(8) Low maintenance requirements.

With the above benefits, electrostatic precipitators have been used industrially for almost a century for the collection of particulate materials. The initial plants were installed to recover valuable material that would have been otherwise lost to the atmosphere, rather than preventing pollution. These plants, mainly in the non-ferrous smelting industry, were designed/costed to produce a 'pay back' in the order of three to five years; consequently collection efficiencies were generally in the order of 90–95 per cent.

For sulphuric acid plants and bullion smelters, because these materials had a higher economic recovery value, even in the early days, design efficiencies of ~99 per cent were typical. Currently, with the tightening of pollution legislation, the primary concern of plant operators is the control of total emission to meet statutory levels, rather than the economic recovery value of the collected material. This is particularly true for the electricity generation industry, where the fly ash has limited usage and, as a result of the large mass flow rates and hence expensive particulate removal plant, the installation and operational costs to comply with legislation have a negative impact on the balance sheet.

The fundamental principle of operation of an electrostatic precipitator is that the particulates are passed through an electric field where they initially receive an electric charge and then, as a charged particle, are deflected across the field to be collected on an earthed plate. Most industrial precipitators are based on a single-stage approach where both charging and migration across the field (precipitation) takes place within the same set of electrodes, as illustrated in Figure 1.1a. Another arrangement, particularly for air cleaning duties, is two-stage precipitation where there is a separate charging field followed by a collection field, as illustrated in Figure 1.1b.

Recent investigations using a hybrid arrangement, comprising a single-stage, then a two-stage precipitator in series, followed by further single-stage fields within the same casing, have indicated that the arrangement produces an increased performance. This improvement is a result of the second pure precipitation field producing an enhanced collection of the agglomerates formed by, and escaping from, the first field [1].

Until the recent development of high frequency direct current supplies (HFDC), which will be detailed later, most currently installed industrial precipitators were electrically energised from mains supply frequency rectified high voltage equipment. This basically comprises a suitably designed and insulated step up transformer, the output from which can be controlled by adjusting the incoming mains supply voltage level. The transformer output is then rectified to produce a high negative direct current (d.c.) voltage. The magnitude of the peak

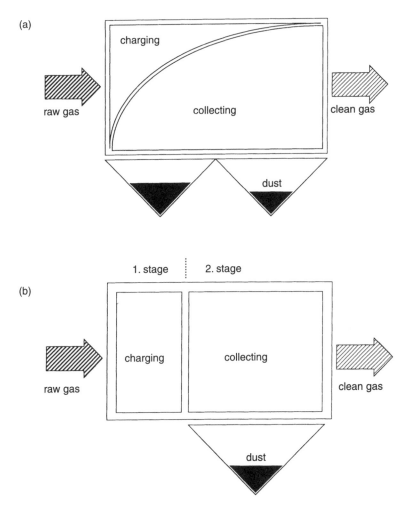

Figure 1.1(a) and (b)   Basic precipitator arrangements

voltage can exceed −100 kV dependent on the precipitator design, collector spacing, etc. Figure 1.2 indicates the basic circuitry of a mains supplied transformer rectifier (TR) set, employing silicon controlled rectifiers (SCR/thyristors) to modulate the incoming supply voltage and silicon rectifiers to produce the d.c. voltage and charging currents.

For air cleaning duties, although energisation is normally derived from the mains supply, the voltages are much lower, 6–12 kV, and typically positive, because of a lower ozone production resulting from the reaction between the ions and the oxygen in the air. This feature will be discussed in detail later.

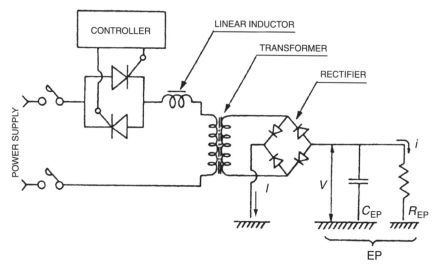

Figure 1.2   Typical arrangement of a mains supply transformer rectifier set

1.2 Arrangement and basic operation of a precipitator

In practice, for large gas flow rates, the single-stage precipitator normally takes the form of a series of vertical parallel plates usually termed receiving or collecting electrodes, which for convenience are normally at earth potential, having discharging elements positioned midway between the plates and insulated from them. The discharging elements, ranging from 3 mm round wire through to specially designed spiked forms, dependent on the supplier's preference and design [2], are energised electrically at a mean d.c. voltage of around −70 kV, for a 400 mm collector electrode spacing.

The high electrical field adjacent to the small diameter electrode elements ionises the gas molecules forming both positive and negative ions. The positive ions are immediately captured by the negatively charged electrodes, while the negative ions and any electrons, generally referred to as a corona discharge, migrate under the influence of the electric field into the inter-electrode space.

As the gas borne particles pass through the inter-electrode space, the larger particles receive an electric charge either by collision with the ions/electrons or by induction charging for the smallest particles. The charged particles then move under the influence of the electric field and migrate to the collecting electrodes, where the charge subsequently leaks away to earth. In the interim, before becoming neutral, the charged particles are retained on the collector surface by a combination of van der Waals' and electric forces.

In a dry precipitator, to ensure that the collection process is continuous, after a period of time the collector electrodes are generally mechanically rapped to remove the deposited material. This time aspect, together with the frequency and intensity of the rapping, is important to minimise rapping re-entrainment and hence maximise collection efficiency. The rapping should be such as to shear the deposited dust from the surface in an agglomerated form, large enough to fall through the gas flow into receiving hoppers or troughs situated beneath the collector plates, rather than exploding the layer from the collection surface, which would result in severe particle re-entrainment.

While the corona wind emanating from the discharge electrodes, coupled with incipient flashover, should assist in keeping the discharge electrodes free of dust, it is normal practice either to rap or vibrate the system to ensure optimum corona production. Figure 1.3 illustrates the major features and equipment arrangement for one form of dry precipitator having horizontal gas flow.

In practice, the advantage of the horizontal flow unit is that, to attain the high efficiencies demanded to meet current legislation, several separate stages or fields can be readily arranged in series. The number of series fields supplied in practice depends not only on the overall efficiency required, but on the supplier in terms of his preferred field length, etc.; in some cases up to nine separately energised fields have been used. In general terms, for efficiencies above 99 per

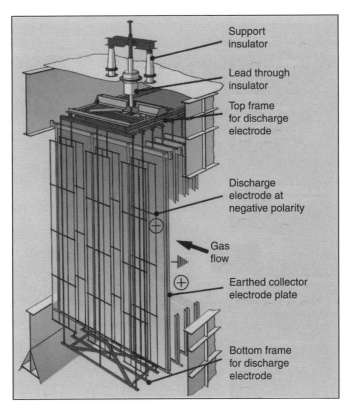

Figure 1.3   Typical mechanical arrangement of a dry precipitator (courtesy Lodge Sturtevant Ltd)

cent, three field arrangements are typical for most installations and for efficiency requirements above 99.5 per cent, four or more fields are to be found.

A typical large power plant dry electrostatic precipitator installation is presented in Figure 1.4. This shows a photograph of two 700 MW lignite fired boiler units, one still under construction, where each boiler unit has a multifield horizontal flow precipitator comprising four separate parallel flows, located downstream of the boiler air heaters to remove the fly ash with a minimum collection efficiency of 99.8 per cent.

A wet form of precipitator is used in a number of applications, where the gases require to be cooled, either to condense additional material, or to assist in particulate precipitation. Alternatively, if the precipitated material is sticky, or in the form of liquid droplets, then the resulting layer can be satisfactorily removed from the internals either by intermittent water washing, or the precipitator is designed such that the internals are continuously irrigated in order to minimise the impact on precipitator operating conditions. Figure 1.5 illustrates a typical arrangement of a wet precipitator having continuous spray irrigation for removing collected deposits as opposed to the mechanical impact rapping shown in Figure 1.3.

Figure 1.4   Photograph of large power plant installation (courtesy Lodge Sturtevant Ltd)

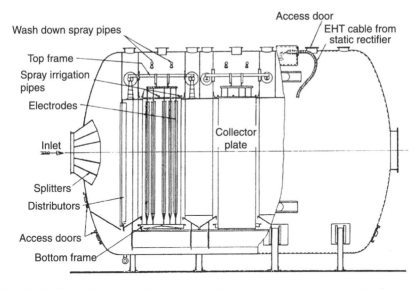

Figure 1.5 *Typical mechanical arrangement of a wet precipitator (courtesy Lodge Sturtevant Ltd)*

Figure 1.6 *Photograph of wet precipitator treating the waste gases from an iron sintering machine following an acid gas scrubber (courtesy Lodge Sturtevant Ltd)*

A typical wet precipitator installation is shown in Figure 1.6. This horizontal flow precipitator is of the wash down type, for final particulate removal following an acid gas scrubbing system on an iron works sintering machine. The collectors of this particular installation are fabricated from carbon fibre coated GRP and the discharge elements are titanium strips, housed in a GRP coated carbon steel casing. This complex and expensive design is required to withstand corrosion arising from the waste gases at the water dew point temperature.

An alternative form of precipitator arrangement is where the collectors take the form of vertical tubes with the discharge electrodes being insulated and hanging co-axially with the tube centre line. To facilitate removal/extraction of the deposited material the tubes hang in a vertical plane, usually with the gas flowing upwards through the tubes. A typical arrangement of a vertical flow tubular unit for removal of self-draining tar particles from a coke oven battery is illustrated in Figure 1.7.

Whereas Figure 1.7 illustrates the internal arrangement of a self-draining mist precipitator, Figure 1.8 is a photograph of an actual installation treating the gases from a coke oven battery. Again, similar to the horizontal flow blast furnace installation shown in Figure 1.5, this mist precipitator installation is to remove tar droplets from the coke oven gases prior to combustion elsewhere on the site. Since the gases are basically devoid of oxygen and hence relatively corrosion free, normal but thicker carbon steel components are used in the construction of this form of unit.

As a result of a more uniform electric field distribution over the whole inter-electrode area, the tubular approach offers an advantage of enhanced precipitation performance; however, because only the inner surface of the tube is used, the construction tends to be more costly than the plate approach where both sides of the plate are used for particle deposition. The tubular design also results in a poorer packing density within the casing, which again impacts on the economics of this approach.

Vertical flow precipitators are now mainly used for self-draining liquid and mist type particulates, where efficiencies in excess of some 99 per cent can be readily achieved from a single field. In the past, when required efficiencies were not so stringent, a large number of 'dry' applications were serviced by vertical flow approaches, these typically having mechanical impact rappers on the collectors and vibrators on the discharge electrodes to remove deposited material.

A practical disadvantage of the vertical flow approach is that unless the required efficiency can be satisfied by a single unit, the gas has to be collected and redistributed into subsequent series fields and as such complicates the site arrangement and increases the cost of the overall installation.

To minimise the cost disadvantage of using round tubes, some designs have adopted either a hexagonal or square 'tube' configuration, fabricated from folded sheet material, while others used a concentric ring (tube) approach to overcome the problem of using only one side of the collector. The difficulty, however, of connecting units in series, if very high efficiencies are required, still poses a costly problem with this approach.

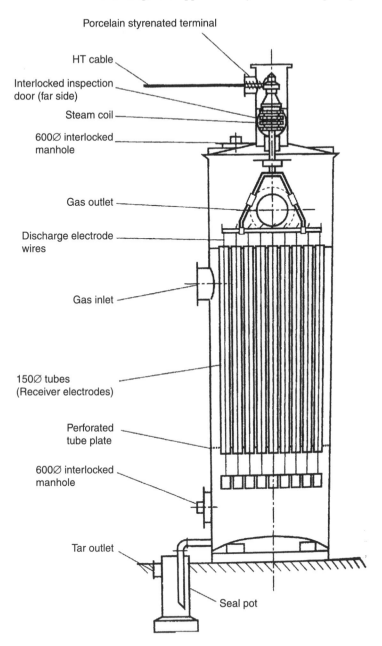

Porcelain styrenated terminal

HT cable

Interlocked inspection door (far side)

Steam coil

600⌀ interlocked manhole

Gas outlet

Discharge electrode wires

Gas inlet

150⌀ tubes (Receiver electrodes)

Perforated tube plate

600⌀ interlocked manhole

Tar outlet

Seal pot

Figure 1.7   Typical arrangement of a detarring mist precipitator (courtesy Lodge Sturtevant Ltd)

Figure 1.8   Photograph of detarring precipitator on a coke oven battery (courtesy Lodge Sturtevant Ltd)

1.3 Versatility

The use and versatility of the electrostatic precipitator can be appreciated from the following list of applications, where different operational designs and arrangements are to be found.

(1) Steam raising – utility and industrial boilers, firing, for example, the following.

   Anthracite, bituminous, sub-bituminous, lignite and brown coals.
   Heavy and light fuel oils, orimulsion, petroleum coke, etc.
   Biomass – wood, straw, chicken litter, grain husks, etc.

(2) Iron and steel manufacturing.

   Blast furnace gas cleaning (basic iron and ferromanganese units).
   Converters, ladles and electric arc furnaces.
   Scarfing and deseaming machines.
   Sinter plants, iron ore pelletisers.
   Foundry and cupola applications.

(3) Metallurgical process plant.

> Dryers, smelters, roasters and refining on non-ferrous plants, e.g. copper, lead, nickel, zinc, etc.
>
> Gold and silver bullion refining operations.

(4) Coal and gas operations.

> Coal drying, carbonisation and treatment, milling and grinding.
>
> Detarring coal gasification and distillation processes.

(5) Cement and lime manufacture.

> Rotary kilns – wet, semi-dry and dry processing plants.
>
> Clinker coolers, limestone crushing.
>
> Raw meal mills, grinders and general feed preparation units.

(6) Waste incineration.

> Municipal, chemical, clinical and hazardous waste disposal units.
>
> Sewage sludge fired installations.

(7) Pulp and paper manufacture.

> Black liquor and Kraft recovery furnaces.

(8) Chemical processing and production.

> Dust and fume from roasters, crushers and dryers,
>
> Fumes arising during the production of fine chemicals and dyes.
>
> Mist collection from sulphuric and phosphoric acid plants.
>
> Gas cleaning in front of acid production units.

(9) Miscellaneous applications.

> Carbon black collection.
>
> Catalyst recovery on refinery 'cat crackers'.
>
> Penultimate gas cleaning stage on uranium reprocessing plant.
>
> Capture and recovery of oil mists.
>
> 'Clean room' applications.

The foregoing summary is not intended to cover all possible applications but relates to many common industrial processes. As each application produces different waste gas flows, temperatures and particulate inlet loading, etc., the precipitator size and design tends to be site specific for a given duty. To obtain a better understanding of the various factors that affect the precipitator design and operation, one must examine the characteristics of both the gas and particulates as they are presented to the precipitator.

1.4 Characteristics of the gases as they affect precipitator design

1.4.1 Composition

The main requirement of the gas carrying the particulates is that it must be capable of maintaining as high an electric field as possible and permit the flow of corona current. This is generally the situation with most industrial process plant.

The composition of some gases can, however, affect the electric operating conditions of the precipitator, which will be discussed more fully later, but generally the corona characteristics are modified by the presence of electropositive or electronegative gases, that is, gases that readily absorb or reject negative ions.

Although one normally thinks of electrostatic precipitators only being applied for the removal of particulates from gas streams, the process has been satisfactorily applied for the removal of particulates from oil streams, again the main criterion is that the oil must be sufficiently insulating to maintain an electric field. The arrangement of the plant is somewhat different to a conventional gas application because the electrodes are normally perforated and the oil passes freely through them, the particles being contact charged before migrating to the area of maximum field intensity between the electrode edge and casing wall. Here they are allowed to agglomerate to a sufficient size to fall out under gravity into the base of the processing vessel from where they can be removed. Some installations use high voltage a.c. for initial treatment, followed by high voltage d.c. for final cleaning of the oil stream; others use d.c. treatment only. Because of the rather specialised nature of this processing, apart from the above mention, the approach will not be further addressed.

1.4.2 Temperature

An advantage of the dry type of precipitator is that the gas in most industrial applications can be delivered directly to the precipitator from the process without the need for additional cooling or treatment. This means that the collected material is usefully captured in a dry state for subsequent disposal/reuse and that when the cleaned gases are emitted from the chimney they are very buoyant and usually at a temperature high enough to result in a steam free discharge.

The gas temperature mainly impacts on the materials of construction; ordinary carbon steel is adopted as the cost-effective material for most applications, which limits the operating temperature up to some 400 °C. In the case of plants operating close to the acid dew point temperature, particularly for wet type environments, the choice of material used for fabrication and long term availability must take corrosion into account [3].

At the other end of the temperature scale, say above some 500 °C, apart from the choice of a temperature resistant material, such as the use of stainless or high nickel alloys, since the kinetic energy of the gas molecules increases with temperature, this results in a significant increase in the electrical discharge characteristics of the gas, which needs to be addressed at the design stage in terms of discharge electrode format and transformer rectifier set sizing.

For gases containing a proportion of electropositive gases, such as the halogens or sulphur hexafluoride, these molecules reduce the impact of the increased molecular kinetic energy of the gases and, if their presence is known at the design stage, this knowledge can assist in the design of the discharge electrode system and power supply requirements.

In practice, precipitators have been installed on various processes where the temperatures range from ambient up to 850 °C.

1.4.3 Pressure

For many of the applications, because most process plants operate close to ambient pressure conditions, gas pressure is not a major effect to be addressed except to ensure that the casing is 'gas tight' and will withstand the operating conditions, that is, either to prevent the egress of process gas or the inleakage of ambient air.

In the case of high pressure, high temperature combustion processes, the pressure, some 7 bar, usually implies that the precipitator casing is constructed in the form of a circular or ovoid cross section. At these pressures, the effect of the increase in kinetic energy of the gas molecules owing to the high temperature tends to be mitigated since the high pressure reduces the higher corona current. Extraction of fly ash from these pressure casings presents a major design difficulty and, normally, several stages of pressure reduction are used to satisfactorily remove the ash, while maintaining the operating pressure within the vessel.

1.4.4 Gas flow rate

In the initial sizing of a precipitator for a given application, although it is important to have an accurate knowledge of the total gas flow such that the correct contact time can be assessed to meet the required efficiency, there is also an optimum gas velocity to be considered.

This optimum gas velocity, which will be addressed later, is determined to some extent by the fly ash characteristics; too high a design velocity for a dry precipitator application can result in particle scouring and potential rapping re-entrainment, while too low a velocity will detract from the overall collection efficiency, since the deposition along the collector plate, most being collected close to the inlet, adversely distorts the electric precipitation field.

It is important that, in order to optimise the collection efficiency, the gas distribution across the frontal area of the precipitator must be as uniform as possible. To produce a typical operational gas velocity of around 1.5 m s^{-1}, which has been decelerated from some 15 m s^{-1} in the inlet approach ductwork, is not an easy task and an acceptable standard of distribution is an r.m.s. deviation of 15 per cent [4]. This standard can be achieved through field corrective testing, large scale model tests or more recently by computational fluid dynamic (CFD) approaches [5].

Although satisfactory field corrective testing can be carried out on smaller plants, on a large electricity generation unit, one must take into consideration that the ductwork can be some 5–6 m high and each corrective splitter can weigh upwards of 500 kg. This makes field corrective work rather impractical, hence the trend, initially toward large scale (>1/10th scale) modelling and more recently CFD approaches, since any necessary splittering, etc., can be established and fitted under controlled fabrication conditions.

Some instances of recent investigations have shown that there is an advantage to be gained from a 'skewed' gas distribution [6]. In this case it has been found that by having a higher gas velocity towards the bottom of the first field and a lower velocity in the bottom of the last field, some of the rapping re-entrainment effects can be partly eliminated, leading to a higher overall effective capture. The theory is that, in the inlet field, the higher velocity and therefore larger mass deposition rate towards the bottom of the collector when rapped, has a shorter distance to fall to reach the hoppers and therefore there is less chance of re-entrainment occurring. In the outlet field, the lower velocity at the bottom of the field provides the opportunity for sheared agglomerates to reach the hoppers without being re-entrained by the horizontal gas stream.

The initial investigations were carried out on boiler plants where the fitting of low NO_x combustion burners resulted in a high carbon carryover in the dust and the 'skewed' distribution assisted in the capture of the large low mass carbon particles. Since then, the process has been adopted to enhance the performance on other plants not necessarily associated with high carbon carryover [7].

Generally, with wet or mist applications, because the deposited material is retained on a wetted surface, operational gas velocities can be appreciably higher than for a dry precipitator, since the risk of particle scouring is significantly reduced.

Although the main application is for processes producing large gas flow rates, for example power generators, the electrostatic precipitator, being extremely versatile, can be used for specialist applications, such as in the pre-treatment of nuclear waste recovery processing plant having gas flows of a few cubic metres per minute.

1.4.5 Viscosity and density

These parameters are determined by the composition and temperature of the gases and affect the precipitation process as follows:

(a) as the charged particles migrate through the inter-electrode space, they are retarded by the effects of gas viscosity and density, and

(b) as the temperature rises, although the gas density decreases assisting ion movement, the viscosity rises and hinders the particle transportation mechanism.

The relationship between these characteristics and precipitator performance will be addressed in greater detail in Chapter 2.

1.5 Characteristics of the suspended material and their possible impact on precipitator performance

1.5.1 Concentration

For many applications the main effect of inlet fly ash concentration is minimal and normally only impacts on the overall removal efficiency requirement.

Generally, increases in the inlet loading over that specified tend to arise from an 'upset condition' on the process plant and are normally associated with the carryover of larger sized particles, which are more readily precipitated than the finer ones. Nevertheless, it is possible that the 'upset condition' can modify other characteristics of the gas and result in an unacceptable emission transgression. Although these upset conditions are generally of a transitory nature, they need to be addressed if the plant has to comply with strict legislative emissions.

1.5.2 Composition and electrical resistivity

The major impact of particle composition and precipitator performance is related to the electrical resistivity of the deposited material. The value of the electrical resistivity of the deposited particles can result in the required exposure/contact time within the precipitation fields varying by a factor of four or more.

For resistivities greater than some 10^{12} Ωcm a phenomenon known as back or reverse ionisation arises, which severely detracts from the overall efficiency. Because of the increasing resistance of the deposited material as the precipitation process proceeds, particles subsequently arriving only slowly lose their charge; consequently, a voltage begins to build up on the deposit and in the worst case reaches a point where positive ions begin to be emitted from the surface of the layer. These positive ions not only neutralise any negative charge on the arriving particles but also considerably modify the electric field such that the overall efficiency is compromised. Measurements have shown that increases in emission by a factor of 10 are not uncommon when faced with this situation. The onset of reverse ionisation can be readily recognised by a significant increase in current flow and an apparent fall in average operating voltage.

Some installations, when faced with precipitator performance difficulties associated with reverse ionisation phenomena, rather than modifying the precipitator in terms of sizing, applying flue gas conditioning or resorting to pulse energisation, have opted for bag filter retrofitting of the casings. This approach has proved successful in a number of installations, particularly in Australia and South Africa, where emissions of <50 mg Nm^{-3} have been regularly attained. Generally, however, unless the power plant is likely continuously to experience the high resistivity problem, the limited choice of suitable filtering media, coupled with its longevity plus a higher operating cost, tends to favour the electrostatic precipitator.

With fly ash having a slightly lower dust resistivity, an alternative operational condition arises, where instead of the voltage continuing to build up on the surface of the deposited layer to initiate positive ion flow, it reaches a value where electrical breakdown occurs within the deposited material, producing a 'leader' which results in flashover across the inter-electrode area. This condition produces an electrical condition typified by a slightly reduced operating voltage with a significant fall in corona current, any attempt to increase the current into the plant only resulting in further sparking and a decreasing performance level.

With conductive materials having electrical resistivities below some 10^8 Ωcm, such as unburnt carbon or metallic particles, although the particles are readily charged by the corona, as they reach the collector they lose their charge so rapidly as to be repelled back into the gas stream. Although the charging and repelling process can occur several times during their transit through the precipitator, some particles can leave the plant without being captured.

Another important factor regarding chemical composition and format is to know if the material is likely to produce a sticky deposit, or is in a liquid phase, when it reaches the collector electrode, such that the correct type of precipitator can be supplied. A further difficulty, usually associated with material composition, is that of cohesive strength of the deposited layer. If the material lacks cohesion and is only lightly bonded together it is likely to be readily re-entrained during rapping of a dry plant and thus detract from the overall performance.

1.5.3 Particle sizing

The electrostatic precipitator can effectively collect particles having diameters from 0.01 μm through to some 100 μm (1 μm (micrometre or micron) = 1×10^{-6} m). The fractional efficiency however, is not constant and generally takes the form of the graph shown in Figure 1.9.

This curve indicates that there is a higher penetration window in the 0.5 μm diameter range, which, as will be shown, coincides with the change from collision to induction charging of the particles. This in itself, provided the particle sizing is available for a specific application/duty, means that the precipitator size needs to be increased to cater for particles falling in this penetration window in order that the plant will meet the required performance level.

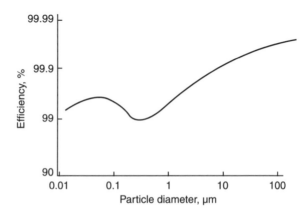

Figure 1.9　Precipitator fractional efficiency relationship

1.5.4 Particle shape

For most applications the particle shape can be either granular, resulting from the comminution of the feed material by grinding, milling, etc., or after being subject to high temperatures, fused into a spherical form. Where fumes are involved, these usually result from the material being initially volatilised in the high temperature zone of the process to subsequently condense into a fume upon cooling. These forms are shown in Figure 1.10 for a fly ash arising from coal combustion [8]. Neither of these forms, provided they are known or assessed beforehand, seriously impact on the overall performance, as long as the size aspect has been considered in the design sizing parameters. For those processes producing a large mass of fine fume, space charge effects, as will be

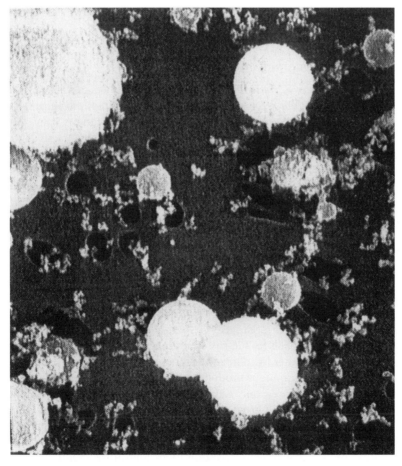

Figure 1.10   Micrograph of fly ash illustrating both granular and fume forms [8]

examined later, can limit corona formation and hence performance unless high emission type discharge elements are employed.

The shapes that can seriously impact on performance are usually those having a thread like form, for example from glass wool product manufacture, where the precipitated fibres can link together and result in short circuiting of the electrical supply. A similar particle linking phenomenon can sometimes arise with platelet type materials, which have a fairly large surface area but virtually no thickness; these are very light in terms of mass and tend to attach themselves end on to the collectors and each other.

Another particle form, which can detract from performance, is the presence or carryover of siliceous cenospheres, which sometimes arise during combustion. These cenospheres are understood to result from carbon particles being trapped within a larger fused siliceous particle, the carbon then burning to produce sufficient gas to inflate the molten mass into a hollow sphere. This problem is not because of their shape, which is usually spherical, but because, in practice, cenospheres are hollow particles having little mass and poor cohesive properties, so can be readily re-entrained by rapping after being deposited.

1.5.5 Particulate surface properties

As far as precipitator performance is concerned, the surface condition/properties of the particle are generally more important than the chemical composition of the actual matrix. In many combustion applications using sulphur bearing carbonaceous fuels, the waste gases can pass through an acid dew point temperature on cooling and any sulphuric acid mist produced subsequently uses the particles as condensation nuclei to deposit a thin layer of highly conductive material on their surface. At higher temperatures of operation, e.g. above 300 °C, there is little possibility of reaching any dew point temperature and it is the chemical matrix of the particle that primarily governs the electrical resistivity.

Where high resistivity particles are met in practice, it will be shown that the electrical operating characteristics and the difficulties of the reverse ionisation phenomenon can be mitigated by the injection of chemical reagents into the flue gases ahead of the precipitator [9]. These reagents typically use the particles as condensation nuclei, depositing a layer of conductive material on the surface of the particles to lower their resistance.

An alternative approach to flue gas conditioning is to modify the method of electrical operation, either by intermittent energisation or by pulse charging. In these, instead of the continuous production of corona, the precipitator is pulsed, such that ions are produced only during the application of the voltage, thereby enabling the charge on the particles reaching the collector to discharge prior to the next set of ions arriving. The capacitive component of the precipitator maintains the field voltage high enough to precipitate particles during the non-pulsing phase. The systems of intermittent and pulse energisation will be discussed in detail in Chapter 7.

1.6 Sizing of electrostatic precipitators

Although the electrostatic precipitator has been used for collecting materials for almost a century, the basic sizing to meet a specific duty still cannot be determined from first principles. The development of computerised programmes for resolving Poisson and Laplace equations has, however, greatly assisted in gaining a better understanding of the physics of precipitation. To date, however, no computer program has been developed for determining the initial design/sizing parameters that will ensure that the precipitator offered will meet the required performance at an economic cost. In practice, most precipitators are sized from interpretation of test data from precipitation plant operating on similar duties, as will be explained Chapter 2.

The physical size and arrangement of an electrostatic precipitator in practice depends on the actual process, the gas and particulate parameters and the efficiency requirement. For a 'difficult' high resistivity dust, for example, that arising from a high ash, low sulphur content coal used on a power plant application, the required precipitator contact or treatment time can be at least 30 s. This compares with a contact or treatment time of less than 10 s to produce the same efficiency for conditioned and easily precipitated dust. While the precipitators will have roughly the same cross section to achieve the optimum gas velocity, the unit having the longer contact time will be at three times the length of the readily precipitated fly ash unit.

While this chapter has concentrated on providing an overview of the range and applications of industrial electrostatic precipitators, plus a short review how some of the carrier gas and particulate parameters affect the precipitator size and design for a specific duty, the following chapter will address the fundamental operational and theoretical characteristics of precipitation and how each phase interacts.

1.7 References

1 WETGEN, D., *et al.*: 'Enhanced Fine Particle Control by Agglomeration'. Proceedings Power-Gen 98, Orlando, USA, 1998
2 GRIECO, G. J.: 'Electrostatic Precipitator Electrode Geometry and Collector Plate Spacing – Selection Considerations'. Proceedings Power-Gen 94, Orlando, Fl., USA, 1994, Paper TP 94-58
3 PARKER, K. R.: 'WESPS and Fine Particle Collection'. Proceedings 8th ICESP Conference, Birmingham, USA, 2001, Paper B5-1
4 ICAC Publication No. EP 7: 'Gas Flow Model Studies'. Institute of Clean Air Companies, 1993
5 SCHWAB, M. J., *et al.*: 'Numerical Design Method for Improving Gas Distribution within Electrostatic Precipitators'. Proceedings of the American Power Conference 56th Annual Meeting, Chicago, USA, 1994
6 ANDERSSON, C., and LIND, L.: 'A Critical Review on the Concept of Gas

Distribution'. Proceedings 8th ICESP Conference, Birmingham, USA, 2001, Paper A 5-1

7 LUND, C., *et al.*: 'Dual FGC Solving ESP Performance Problems'. Proceedings 7th ICESP Conference, Kyongju, Korea, 1998, pp. 550–60

8 KAUPPINEN, E. I., *et al.*: 'Fly Ash Formation at PC Boilers Firing South African and Colombian Coals'. Proceedings EPRI/EPA International Conference on *Managing Hazardous and Particulate Air Pollutants*, Toronto, Canada, 1995

9 BUSBY, H. G. T., and DARBY, K.: 'Efficiency of Electrostatic Precipitators as Affected by the Properties and Combustion of Coal'. *J. Inst. Fuel*, 1963, **36**, pp. 184–97

Chapter 2

Fundamental operation of an electrostatic precipitator

2.1 Introduction

Although the initial investigators, such as Lodge [1], employed an electrostatic generator such as a Wimshurst Machine successfully to demonstrate the precipitation phenomenon, once the development was applied in the field, it became apparent that more electrical power, in terms of current, was necessary to ensure satisfactory particle charging such that the particles could be continuously precipitated. More correctly the process should be referred to as electrical precipitation, because with modern installations one can find rectifier equipment having outputs up to 200 kW and for large power station applications a power consumption up to 2 MW.

The basic principle of operation of an electrostatic precipitator, as already indicated, is that the gas borne particles are passed through an electric field where they are initially charged by means of a corona discharge and then, as a charged particle, are deflected across the electric field to migrate and be deposited on the collection or receiving electrodes.

There are two basic electrostatic precipitator arrangements found in practice: a single-stage approach, where both charging and migration arise using a single set of electrodes; and the two-stage approach, where separate charging and precipitation fields are involved.

The fundamentals for either arrangement can be summarised as:

- the production of a corona field to create ions,
- the charging of the particles by the ions,
- the migration of the charged particles through the field,
- the arrival of the charged particle at the receiving electrode,
- the removal of the deposited particles into the hoppers.

2.2 Ion production

Although there are various methods of particle charging, for example tribo-electric, ultraviolet and radiation effects, for industrial precipitator applications corona charging is universally used as being the most efficient and cost-effective approach. The physics of corona or ion production therefore occupies an essential position in the practice of electrostatic precipitation. Investigations into the physics of ionisation date back to the middle of the nineteenth century and are still ongoing, particularly since the development and application of computational fluid dynamics (CFD) using fast computers capable of deriving satisfactory solutions to Poisson and Laplace equations.

The early investigations were concerned with developing voltage current relationships of corona discharge and the effects of different means of electrical energisation. In 1862, Gaugain [2], working with concentric cylinders, found that for a given outer diameter electrode, the electrical breakdown voltage was primarily dependent on the radius of the central emitter electrode.

Qualitatively, the following equation represents the breakdown potential of such a system, where the radius of the inner electrode, r, is very much smaller than R, the radius of the outer passive electrode, i.e. $r << R$:

$$E = A + C/r^{1/3}, \tag{2.1}$$

where E is the electrical breakdown field, r is the radius of the inner electrode, and A and C are experimental constants.

Röentgen [3] in 1878, working with point/plane electrodes, found that a certain voltage had to be applied to initiate a corona current flow. This corona onset voltage was dependent on the sharpness of the point, gas pressure and polarity of the point electrode. Further investigations produced a parabolic relationship for negative corona current flow as:

$$I = AV(V - M), \tag{2.2}$$

where I is the corona current flow, V is the applied voltage and A and M are experimental constants.

The next significant finding, as regards conventional early concentric wire and tube precipitator arrangements, can be attributed to work carried out by Townsend [4]. It was noted that with negative ionisation of the central discharge element, with the outer being connected to the positive terminal of the supply voltage, the appearance of the corona discharge was very different to when the inner electrode was positively energised. In the case of negative energisation the corona appears as bright flares which tend to move across the surface of the electrode, whereas with positive energisation the corona appears as a diffuse glow surrounding the electrode. Other differences found were that with negative energisation the corona had a distinctive hissing sound, the corona initiation voltage was lower and the breakdown potential higher.

Figure 2.1 indicates the difference in electrical characteristics of the two forms of energisation for the same electrode arrangement.

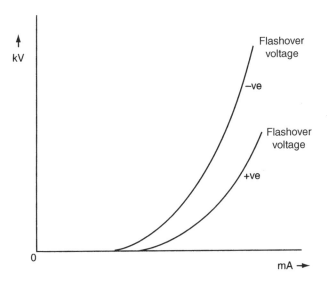

Figure 2.1   Comparison of discharge characteristics for negative and positive energisation of a wire–tube electrode arrangement

In general, these findings probably explain why negative energisation is normally used for industrial gas cleaning operations, although for air cleaner type duties, because of a lower ozone production rate, positive energisation is generally used.

The conduction of electricity through gases is fundamentally different to that in solids and liquids, which contain 'charge carriers' that move under the influence of the electric field to produce a current flow. For metallic materials, the current arises from the movement of electrons that migrate through the crystal lattice; for semiconductors, the charge carriers can be either electrons or 'holes' that migrate through the material; for insulating materials, e.g. silica and alumina, ionic conduction only occurs at high temperature, when the mean free electron path is significantly increased, and in the case of conductive liquids, current flow results from electrolytic ion conduction through the liquid. Whereas with gases, ions need to be provided, or produced from some outside force or agency, to induce a current flow that will cross the inter-electrode space.

For electrostatic precipitators applied to gaseous applications, the major outside agency for ion production is by high voltage electric input. Figure 2.2 indicates a typical electric field distribution across the inter-electrode area, between a small diameter emitter and a much larger passive electrode, i.e. $r << R$, for a tubular type of precipitator.

It will be noted that the electric field adjacent to the emitter is extremely high and it is this electrical stress which excites any free electrons in the immediate vicinity. These fast moving electrons acquire sufficient energy from the applied

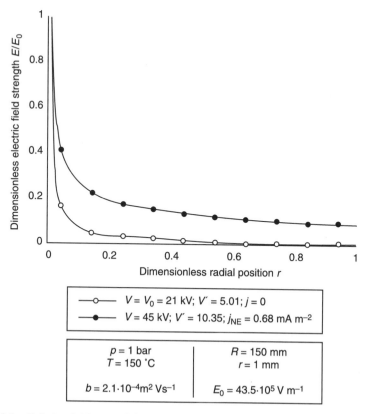

Figure 2.2   Relative field strength between the electrodes of a tubular precipitator
arrangement

electric field to collide with other gas molecules to produce further free electrons
and positive ions. Townsend, working in this area, proposed the concept of
a chain reaction or electron avalanche, in which each new electron produced
generates new electrons by ionisation in ever increasing numbers.

The number of ions at a distance x from the active zone can be represented by

$$n = n_0 e^{ax}, \tag{2.3}$$

where n_0 represents the number of ions at a distance $x = 0$ and a is the Townsend
ionisation coefficient, which varies with the gas temperature, pressure and
resultant field strength.

In general terms, where the field varies with distance across the field x, as in a
precipitator, then the equation takes the integral form

$$n = n_0 e^{\int_0^x a\,dx} \tag{2.4}$$

(note the term $1/a$ is the mean free path of the electron between collisions).

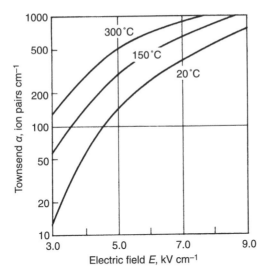

Figure 2.3   Effect of electric field strength and temperature on ion pair production

Figure 2.3 indicates the relationship between the Townsend ionisation coefficient (ion pair production) for air at atmospheric pressure and at different temperatures and field strengths.

Although few industrial gases comprise pure air (79 per cant $(N_2 + CO_2) + 21$ per cent O_2) as the carrier gas, the curve indicates the effect of increasing field strengths on the ionisation characteristics. At 20°C, doubling of the field strength results in the number of ion pairs, or Townsend coefficient, increasing by a factor of 20. The impact of a rise in gas temperature producing significantly increased ionisation because of the larger mean free electron paths is also indicated.

In practice, with the system negatively energised, although there are a large number of ion pairs immediately adjacent to the discharge element, as the electrons rapidly move across the field area they collide with and attach themselves to gaseous molecules to produce negative ions, while the positive ions are attracted towards the discharge element and, although initially during transit producing further ion pairs, on reaching the element take no further part in the process.

Adjacent to the discharge element there is an abundance of electrons, but as the distance from the element increases, because of attachment, the number of electrons decrease and there is a corresponding increase in the number of negative ions. The net number of electrons at a distance x from the electrode is represented by the following equation:

$$n = n_0 e^{\int_0^x (\alpha - \eta)\, dx} \tag{2.5}$$

where η = coefficient of attachment or the Townsend second coefficient.

Note. Those gases in the right hand region of the Periodic Table, such as Cl_2, HF, O_2, SO_2 and SF_6, all being deficient in electrons in their outer shell, have a great affinity for electron attachment and are termed electronegative gases. Their affinity for electron attachment, even if the gases are present in minor quantities, tends to reduce ionisation and suppresses the electrical discharge in gases. Electropositive gases such as N_2, H_2, Ar, etc., in high concentrations have little affinity for electron attachment and consequently do not produce negative ions and result in little current flow up to their breakdown voltage.

An example of a uniform stable self-maintaining gas discharge between an emitter and passive or receiving electrode is illustrated in the case of a concentric wire and tube precipitator arrangement, where the radius of the emitter is small compared with that of the passive electrode, as shown in Figure 2.4.

In the case of positive energisation, normally found in air cleaning applications, the primary electrons produced at the boundary of the visible glow are attracted towards the emitter. In moving through the field they collide and produce new ion pairs by impact ionisation, the positive ions migrating towards the passive earthed electrode. Except for the collection of electrons, the emitter itself plays little part in the ionisation phenomena, which is essentially a gas process with the primary electrons being released from the gas molecules through photoelectric effects in the plasma region.

Negative corona is a feature of gases or admixtures that exhibit appreciable electron attachment and as such are sensitive to gas composition and temperature. The corona can range from virtually zero to being a highly stable discharge

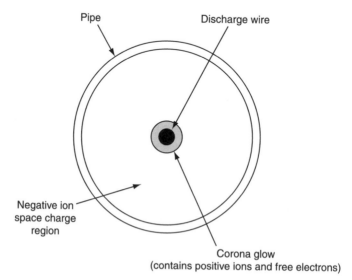

Figure 2.4   Schematic tubular precipitator arrangement indicating active and space charge areas

dependent on operating characteristics. Unlike positive energisation, the emitter plays an important role in the ionisation phenomena; any positive ions generated in the plasma region are attracted towards the emitter because of the high field strength, to produce further electrons on impact.

The ultraviolet radiation in the plasma region releases even further electrons; hence we have an abundance of fast moving electrons in the plasma region. These electrons move rapidly away from the electrode to produce ion pairs on impact with the gas molecules. The emitter immediately captures the positive ions, while the negative ions and any free electrons migrate towards the passive electrode.

As indicated earlier, the visual appearance of negative corona takes the form of tufts or bright glow points or Trichel pulses [5], which on a plain smooth emitter rapidly move about. The number of tufts and their luminescence increases as the energising voltage is raised producing a higher corona current discharge.

2.3 Particle charging

For industrial precipitator applications, where negative ionisation is used because of its lower corona initiation voltage and higher breakdown potential, particle charging occurs in the area between the active plasma region and the passive electrode surface. This area comprises a high space charge having neutral ions, negative ions, plus some free electrons, all moving towards the passive electrode as a result of the electric field.

As the gas borne particles enter the corona derived space charge region of the field, two charging mechanisms occur; the first is by ion attachment, i.e. field or impact charging, and the second by ion diffusion charging. The field or impact charging predominates for particles greater than some 1 μm, whereas diffusion charging is essential for particles less than 0.2 μm in diameter, both processes occurring in the intermediate size range.

While field charging requires the presence of an electric field to drive the free mobile charge carriers, the diffusion process is based on randomly moving gas ions arising through temperature effects as described by the kinetic theory of gases, that is, Brownian motion. It will be shown that Brownian motion plays an important role in the collection of particles in the sub-micrometre size range in spite of their saturation charge being much smaller than for the larger particles.

Over recent years a great deal of numerical modelling work has been carried out using computational fluid dynamics to derive particle charging models [6, 7, 8]; however, the basic models need experimental support since the equations cannot be analytically solved. A reasonable alternative to modelling is the analytical work carried out by Cochet [9], who developed an equation, which appears to give reasonable correlation to actual precipitator measurements in the critical size range, as carried out and reported by Hewitt [10].

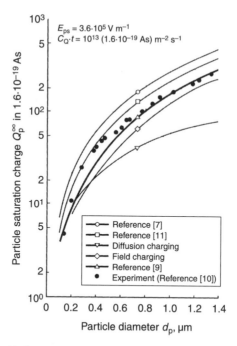

Figure 2.5   Relationship between saturation charge and particle diameter

Figure 2.5 shows a series of curves relating particle saturation charge to its diameter, based on Cochet's equation and experimental results from Hewitt and others in relation to the mathematical modelling carried out by Smith and co-workers [11].

The particle saturation charge, Q^{∞}_{p}, according to Cochet, is given by the following formula

$$Q^{\infty}_{p} = \left\{ (1 + 2\lambda/d_{p})^{2} + (2/1 + 2\lambda/d_{p}) \times \left(\frac{\varepsilon_{r} - 1}{\varepsilon_{r} + 2} \right) \right\} \pi \varepsilon_{0} d_{p}^{2} E \qquad (2.6)$$

where ε_{r} is the electrical permittivity of the particle, ε_{0} is the electrical permittivity of the gas, d_{p} is the particle diameter, E is the electric field strength and λ is the mean free electron path.

In simplified terms the equation can be rewritten

$$Q^{\infty}_{p} = pEd_{p}^{2}, \qquad (2.7)$$

where p is a constant, which varies between one and three for non-conductive particles and is two for conductive particles.

It can be shown that all particles reach around 90 per cent of their full saturation charge in less than 0.1 s. Hence, for all practical conditions, it can be assumed that on entering the precipitation field, having a typical exposure/treatment time of 3–5 s, the particles rapidly achieve their saturation charge.

2.4 Particle migration

Within the precipitation field a particle experiences the following forces acting upon it: a momentum force F_m, an electrical force F_e and a drag force F_d,

where $F_m = m\,a$,
$\quad\quad F_e = Q_p\,E$,
$\quad\quad F_d = Re\,A\,Co$ (*Re* is the Reynolds number, and *Co* is the Cunningham correction factor),

and where $F_m + F_e + F_d = 0$ (steady state condition). (2.8)

Prior to solving this differential equation, the drag force F_d has to be calculated.

In the case of low Reynolds numbers, the drag coefficient is given by

$$\overrightarrow{F_w} = 3\pi\eta d_p|\overrightarrow{v} - \overrightarrow{\omega}| \times \frac{1}{Cu}$$

$$\text{where } Cu = 24/Re_p. \quad\quad (2.9)$$

As the particle size d_p reduces and approaches the region where the fluid loses its continuum (mean free path of molecules $= \lambda$) then Stokes law needs to be corrected by the Cunningham correction factor (*Co*):

$$Co = 1 + 1.246 \times 2\lambda/d_p + 0.42 \times 2\lambda/d_p \times \exp{(-0.87d_p/2\lambda)}. \quad (2.10)$$

This relationship is plotted in Figure 2.6, which shows the significant correction factor that arises for sub-micrometre sized particles.

The drag force (or Stokes law) can be written as

$$F_d = 3\pi\eta d_p\omega_{th} \times 1/Co, \quad\quad (2.11)$$

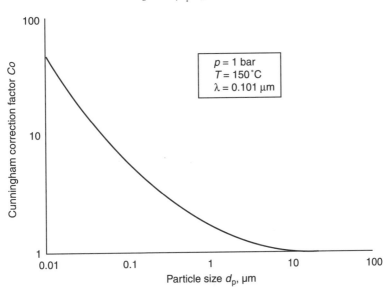

Figure 2.6   Relationship between particle size and the Cunningham correction factor

where η is the gas viscosity, d_p is the particle diameter, Co is the Cunningham correction factor and ω_{th} is the theoretical particle migration velocity.

Assuming the fluid has no component acting towards the passive electrode and all particles achieve their saturation charge, then the equation of motion of a charged spherical particle in an electric field is characterised by

$$\frac{d\omega_{th}}{dt} + \frac{3\pi\eta d_p \omega_{th}}{mCo} = \frac{Q^\infty_p E}{m}. \tag{2.12}$$

Taking $\omega_{th} = 0$ at $t = 0$, the solution to the above equation can be readily found, i.e.

$$\omega_{th} = \frac{Q^\infty_p E}{3\pi\eta d_p} \times Co, \tag{2.13}$$

$$\text{since } Q^\infty_p = pEd_p^2$$

$$\text{then } \omega_{th} = \frac{E^2 \times d_p \times Co}{\eta}. \tag{2.14}$$

The significance of this relationship is as follows.

(a) As the limiting charge on the particle is proportional to the radius squared, theoretically the migration velocity of the particle will increase with particle size.

(b) Because the electric field is proportional to the applied voltage, the theoretical migration velocity is proportional to the voltage squared.

It will be appreciated from this fundamental approach that the operation of an electrostatic precipitator is dependent on having a voltage high enough to produce an electric field in order to precipitate the particles and have sufficient current capability to satisfy ion production for the initial charging of the particles.

Figure 2.7 indicates the relationship between the theoretical migration velocity against particle size for three different field strengths in a typical flue gas at 150 °C, for particles having an electrical permittivity of ten. Although, because of the logarithmic scale, the effect of the field strength squared factor is not apparent, if one examines the actual migration velocity figures, the effect is more pronounced.

Although the foregoing discussion portrays the theoretical approach to the migration of charged particles through an electric field, the derived figures of theoretical migration velocity should not be confused with the 'effective migration velocity', which is derived from plant efficiency measurements and the specific surface area of the precipitator. The 'effective migration velocity' derived from measured efficiencies and the specific collection area for the precipitator should more realistically be considered as a measure of a precipitation performance factor rather than a measure of the average theoretical particle migration velocity.

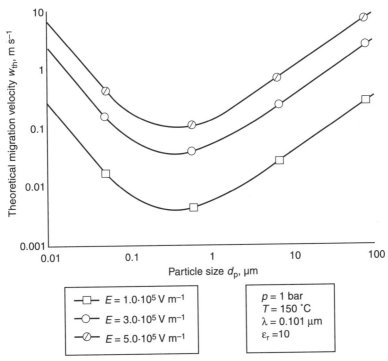

*Figure 2.7   Relationship between the theoretical particle migration velocity and particle
size as derived from the above equations*

Whichever approach is considered, the general trend of a much reduced migration velocity, based on a fractional efficiency, at around 0.5 μm is still apparent. The subsequent increase in efficiency for smaller particles is the result of Brownian motion aiding the charging and transfer of these particles toward the collector plates. In order to derive a fractional efficiency curve as shown in Figure 1.9, it is necessary to determine the particle sizings at both the inlet and outlet of the precipitator and then to evaluate separate grade efficiencies from the overall mass precipitator efficiency. The separate grade efficiencies can then be transposed into effective migration velocities from the Deutsch relationship.

Although the previous discussion describes the motion of a charged particle through an electric field, until the particle has been removed from the collector plate and transferred to the receiving hopper it should not be considered as being collected. This is one of the many reasons why the theoretical values of migration velocity are usually far higher than those based on actual measured efficiency/plate area calculations. The theoretical approach also assumes that the particle loses its charge on arriving at the collector and plays no further part in the process. In practice, as indicated in the previous chapter, the particulates arriving at the collector impact on the electrical operating conditions and hence affect performance.

2.5 Particle deposition and removal from the collector electrodes

As the particle arrives at the collector, the first reaction is that it is held by a combination of van der Waals and electrical attraction forces. Depending on the thickness of the deposited layer and the electric resistance of the layer, the charge on the particle leaks away to earth, leaving the particle to be retained only by van der Waals forces. Unless the deposit is periodically removed, the precipitation process will slowly degrade because the deposit will have a deleterious impact on the electrical operating conditions. In the case of a dry precipitator this is achieved by some form of mechanical rapping and in the case of a wet precipitator, depending on the design, by periodic water washing (flushing), a continuous spray irrigation system, or having a film of water flowing over the collectors (film flow).

For a dry application the frequency and intensity of the rapping must be such as to shear the deposit as a layer from the plates, rather than attempting to disintegrate the layer completely as this would lead to severe rapping re-entrainment. Theoretically it can be shown that to remove a 10 μm particle from a metal surface would require a force of some 1000 g. Although such intensities are achievable it would not only remove the particle, but, in addition to completely disintegrating the deposited layer, intensities of this magnitude would also result in fatigue and ultimate failure of the components being rapped. Ideally for the particles to reach the hopper the effective size of agglomerate to overcome the horizontal component of velocity should be in the order of at least 500 μm diameter [12]. A compromise, therefore, has to be achieved on any operating plant to have sufficient energy imparted on the collectors to remove the deposit in an agglomerated form but not result in mechanical damage to the components.

As the rate of deposition varies along the length of the precipitator, the frequency of rapping reduces from the inlet through to the outlet fields. This enables the deposited layer between blows to achieve a thickness, such that it can be sheared from the plates as an optimum sized agglomerate in order to reach the hoppers with the minimum of re-entrainment. Because the mean particle size of the deposited material decreases along the length of the precipitator, ideally the intensity of the rapping blow should increase as the particle size becomes smaller towards the outlet end of the precipitator because of the increased packing density and cohesion between particles.

Although the majority of the dust is deposited on the collectors, some is deposited on the discharge electrodes and, in order to optimise corona production, the discharge electrodes should be kept deposit free. On a dry plant this is achieved by rapping systems similar to those employed for the collectors; for a wet system, water washing of the internals is usually effective in maintaining corona emission.

The relationship, indicated in Equation (2.14), is important in practice, because to optimise performance it is important that the voltage applied across the electrode system is maintained at the maximum possible level. The actual operating voltage across the electrode system is largely dependent on the properties of the gas, the material suspended in the gas, the layer of deposited material and the design of the electrode system. Although it would be ideal consistently to operate just below the electrode breakdown potential, since the complete system is dynamic and conditions are constantly changing, an automatic voltage control system is used in order to maintain the voltage, and hence performance, as high as possible.

2.6 Precipitator efficiency

As indicated earlier, the fundamental sizing of an electrostatic precipitator for a specific duty cannot at present be derived from first principles. Most equipment suppliers use a 'performance factor' derived from the measurement of efficiency on a plant having similar process characteristics in order to obtain a new size for the required duty.

Deutsch, one of the early theoretical investigators, working in the mid-1920s, proposed [13] that the performance or collection efficiency of a precipitator took the form of an exponential equation:

$$\text{efficiency } \eta = 1 - e^{-a}, \tag{2.15}$$

where a is dependent on factors related to the precipitator design and physical properties of the gas and dust.

In deriving the above relationship, Deutsch assumed that infinite turbulence produced a homogeneous distribution of the particles within the gas stream, the particles were fully charged and the gas velocity was uniform, as was the corona current distribution on the collector plates, none of which is strictly true in practice.

During his investigations, by changing the gas velocity through the precipitator, while keeping all other factors constant, Deutsch derived the following efficiency relationship:

$$\text{efficiency } \eta = 1 - e^{-2l/v}, \tag{2.16}$$

where l is the length of field and v is the gas velocity.

(*Note.* It should be recognised that l/v is the reciprocal of the precipitation contact time.)

The above relationship was later transposed by the precipitation industry and used for many years to relate the size of the precipitator, collection efficiency and gas flow rate for plants operating on similar duties and inlet conditions. Although the formula is often assumed to be theoretically based, it is no more than a useful method of comparing precipitator performance levels.

The relationship is probably more recognisable in the following form:

$$\text{efficiency } \eta = 1 - e^{-A\omega/V}, \tag{2.17}$$

where ω is the effective migration velocity (m s^{-1}), A is the area of the collecting plate (m^2) and V is the gas flow rate (am^3 s^{-1}).

Rearranging,

$$\omega = \log_e [1/(1 - \eta)] \times V/A. \tag{2.18}$$

It is important to note that the value ω, the effective migration velocity in a real precipitator, is not equal to the theoretical value ω_{th}, derived from Equation (2.14),

i.e. Deutsch $\omega \neq \omega_{th}$.

2.7 Practical approach to industrial precipitator sizing

Although the Deutsch relationship was used for many years as a design tool by the precipitation industry for comparing and sizing precipitators to cater for different gas flow rates and efficiencies on various industrial applications, it was not until the 1960s, when difficulties arose in meeting more stringent emission requirements, particularly when dealing with fly ash from low sulphur coals in the electricity generation industry, that modifications to the traditional Deutsch relationship were considered.

One of the better known modifications was that derived by Matts and Ohnfeldt [14], who produced the following relationship, termed the modified Deutsch formula, for deriving an improved working value as an effective migration velocity ω_k:

$$\omega_k = \log_e [1/(1 - \text{efficiency})]^2 \times V/A. \tag{2.19}$$

An alternative approach was that proposed by Petersen [15], who developed the following relationship:

$$1/(1 - \text{efficiency}) = (1 + b \cdot \omega \cdot A/V)^{1/b}, \tag{2.20}$$

where b is an empirically derived constant $= 0.22$, which holds true for both the power and cement industry applications.

In practice, whichever of the above relationships is used, the performance of the resultant installed precipitator invariable meets the required efficiency level.

The practical implication of these modified approaches is that, for a change in efficiency, based on an installed plant whose operational data is known, the required plate area increase is around twice that derived from the straight Deutsch formula.

The application of the modified Deutsch equation will be appreciated from the following example.

For the high efficiencies presently demanded, if one considers a precipitator installation for a 500 MW pulverised coal fired boiler unit, burning a typical UK coal containing 20 per cent ash and having a sulphur content of 1.5 per cent, the precipitator data sheet would read as follows.

DATA SHEET DETAILS

Application – pulverised coal fired boiler, conventional steam turbine electrical generation 500 MW

Coal – HHV ~7000 kcal kg^{-1}, volatile content ~30 per cent, ash content 20 per cent, moisture as received 12 per cent and the sulphur in the coal 1.5 per cent.

Gas conditions at precipitator inlet

Gas volume at air heater outlet 750 am^3 s^{-1}
Gas temperature 125 °C
Oxygen in gas at precipitator ~6 per cent dry basis
Assumed inlet fly ash loading 20 g Nm^{-3} dry
Required emission level 50 mg Nm^{-3} at 6 per cent oxygen, dry basis
Efficiency of collection 99.75 per cent
Number of precipitator casings = 2
Number of series fields per casing = 4
Collector plate size 4000 × 15 000 mm
Collector plate centres = 400 mm
Number of ducts per precipitator casing = 38 (39 collectors)
Total plate area = 2 × 4 × 2 × 4 × 15 × 38 = 36 480 m^2
Average gas velocity through precipitator = 750/2 × 15 × 38 × 0.4 = 1.64 m s^{-1}
Average contact time = 4 × 4/1.64 = 9.72 s
Specific collecting area (SCA) = 48.64 m^2 am^{-3} s
Deutsch effective migration velocity (Equation (2.18)) = 12.32 cm s^{-1}
Modified migration velocity (Matts–Ohnfeldt – Equation (2.19)) = 73.8 cm s^{-1}
Approximate footprint of installation, 35 m wide × 30 m long
Height of installation, ~40 m, dependent of vendors' design, ground clearance, etc.
Number of independent TRs (assume mains rectification) = 16 (2 per field)*
TR rating 110 kV peak at 2000 mA (cap-res)**

* Although the above example proposes that 16 identical TRs would be used to energise the precipitator installation, each rated at 110 kV peak and 2000 mA, by matching the actual power consumption for each field, it may be possible to lower the capital cost of the complete electrical installation by reducing the size and rating of some of the TRs, which would additionally involve smaller and lower cost input power cabling and switch gear. In practice, the economics and ability to interchange rectifier equipment in the event of a failure, usually means that the TRs are all rated for the highest power consumption, typically the outlet field, which has the lowest space charge and hence the maximum power consumption.
** Value depends on type of discharge electrode.

If legislation should change and the required emission fall to 25 mg Nm^{-3}, the efficiency of collection would need to increase to 99.875 per cent and the above precipitator size (plate area or contact time) would have to increase by approximately 25 per cent.

Assuming the modified Deutsch number holds at 73.8 cm s^{-1}, then from Equation (2.19), the plate area A would be given by

$$A = \log_e [(1/ 1 - 0.998\ 75)]^2 \times 750/0.738$$

$$= 45\ 410\ \text{m}^2.$$

The SCA would need to increase to 60.55 m^2/m^3/s and the contact time to 12.2 s, that is, an extra field would be necessary to meet the 25 mg Nm^{-3} emission value.

The foregoing example is included to illustrate the size and complexity of a current precipitator installation as would be applied to a relatively easy application. If the coal had been from another source containing a much lower sulphur content, then as will be shown later, the installation would have to be considerably increased in size, well above that indicated.

Although most of the recent investigations have been carried out in the power generation field, the theory and usage of the modified Deutsch approach has been confirmed on many different processes in order to relate results from older lower efficiency installations to derive size increases to satisfy current emission legislation.

2.8 References

1   LODGE, O. J.: 'The Electrical Deposition of Dust and Smoke with Special Reference to the Collection of Metallic Fume and to Make Possible Purification of the Atmosphere', *J. Soc. Chem. Ind.*, 1886, **5**, pp. 572–6
2   GAUGAIN, J. M.: 'On the Disruptive Discharge', *Annales de Chimie et de Physique*, 1862, **64**, p. 175
3   RÖENTGEN, V.: *Göttinger Nach*, 1878, p. 390
4   TOWNSEND, J. S.: 'Electricity in Gases' (Oxford University Press, 1915)
5   TRICHEL, G. W.: 'The Mechanism of the Negative Point-to-Plane Corona near Onset', *Physical Review*, 1938, **54**, p. 1078
6   MURPHY, A. T., *et al.*: 'A Theoretical Analysis of the Effects of an Electric Field on the Charging of Fine Particles', *Trans AIEE*, 1959, **78**, pp. 318–26
7   LIU, B. Y. H., *et al.*: 'On the Theory of Charging Aerosol Particles in an Electric Field', *Journal of Applied Physics', 1968*, **39**, pp. 1396–402
8   LIU, B. Y. H., *et al.*: 'Combined Field and Diffusion Charging of Aerosols in the Continuum Regime', *Journal of Aerosol Science*, 1978, **9**, pp. 227–42
9   COCHET, R.: 'Lois Charge des Fines Particules (submicroniques) Etudes Théoriques – Controles Récents Spectre de Particules'. Coll. Int. la Physique des Forces Electrostatiques et Leurs Application, Centre National de la Recherche Scientifique, Paris, 1961, **102**, pp. 331–8
10   HEWITT, G. W.: 'The Charging of Small Particles for Electrostatic Precipitation', *Trans. AIEE*, 1957, **76**, p. 300
11   SMITH, W. B., *et al.*: 'Development of Theory for the Charging of Particles by Unipolar Ions', *Journal of Aerosol Science*, 1976, **7**, pp. 151–66
12   BAYLIS, A. P., *et al.*: 'Collecting Electrode Rapping Designed for High

Efficiency Electric Utility Boiler Electrostatic Precipitators', Proc. 4th EPA/ EPRI Symposium on the *Transfer and Utilization of Particulate Control Technology*, Houston, Tx., USA, 1982; EPRI, Palo Alto, Ca., USA

13 DEUTSCH, W.: 'Bewegung und Ladung der Elektrizitätsträger im Zylinderkondensator', *Ann. Phys.*, 1922, **68**, pp. 335–44

14 MATTS, S., *et al.*: 'Efficient Gas Cleaning with SF Electrostatic Precipitators', *Fläkten*, 1963/4, **1–12**, pp. 93–110

15 PETERSEN, H. H.: 'A Precipitator Sizing Formula'. Proceedings 4th ICESP Conference, Beijing, China, 1990; Int. Academic Pub., Beijing, 1993, pp. 330–38

Chapter 3

Factors impinging on design and performance

As will be appreciated from the preceding chapters, there are a large number of factors relating to both the carrier gas and particulates that affect the design, operation and performance of an electrostatic precipitator. This chapter will examine some of these characteristics in greater detail to explain how they impact on the electrostatic precipitator both mechanically and electrically.

3.1 Effect of gas composition

For combustion process applications, the carrier gas typically comprises nitrogen, carbon dioxide, moisture and oxygen, together with trace quantities of other gases, e.g. sulphur dioxide, sulphur trioxide, nitrogen oxides, etc., dependent on the process feedstock. Other processes, such as metallurgical smelters, can have high concentrations of sulphur dioxide, while gases from reduction type processes, for example, cleaning of blast furnace or coke oven gases, have only trace quantities of oxygen, <1 per cent, but can contain hydrogen sulphide, cyanic derivatives and other compounds.

Corona inception, as indicated, occurs when the electric field adjacent to the discharge element reaches a certain gradient irrespective of whether positive or negative energisation is used. The actual corona current flow depends on the energising voltage and whether the gases have electropositive or electronegative characteristics. Gases in the right hand area of the Periodic Table, such as chlorine, oxygen, sulphur dioxide, hydrogen fluoride, etc., are termed electronegative gases, which, being deficient in electrons in the outer shell, have a great affinity for electron attachment to produce negative ions. Other gases such as nitrogen, hydrogen, argon, etc., are termed electropositive gases and have little affinity for electron attachment and consequently do not produce negative electrons in any great quantity.

Although the foregoing is true for pure electropositive gases, the presence of small quantities of electronegative gases, e.g. oxygen or sulphur dioxide, as

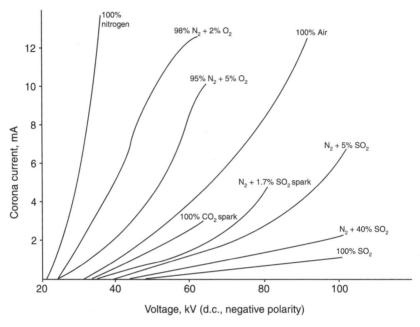

Figure 3.1   Typical corona characteristics of gas mixtures

typically met in practice, significantly modifies the corona characteristics of the mixture, as indicated in Figure 3.1.

From these curves, in the case of pure sulphur dioxide, the number of ions produced is small and consequently results in little corona current flow even up to the breakdown voltage of the system. On the other hand, small quantities of electronegative gases, e.g. oxygen or sulphur dioxide, modify the corona characteristics, enabling the system adequately to charge and hence precipitate any entrained particles.

Because of the importance of these small quantities of electronegative gases on the corona characteristics, it is important that a complete and full gas analysis is available for any application to assess if particle removal by electrostatic precipitation is feasible. For example, it is not economic or practicable to use electrostatic precipitation for final cleaning of nitrogen or argon gases produced from bulk manufacturing plants.

3.2 Impact of gas temperature

The gas temperature plays several important roles in the design and operation of an electrostatic precipitator. The first is that temperature can dictate the materials of construction; for example, for temperatures up to some 400 °C, normal carbon steel is generally used for both containment casing and internal components. For temperatures around 500 °C, some form of stainless or alloyed steel would be used, while for higher temperatures, one would consider high nickel alloys for construction purposes. For the larger installations a further difficulty facing the precipitator designer is that of differential thermal expansion. This needs to be taken into account, particularly for plants designed to operate at elevated temperatures to avoid distortion leading to degraded precipitator performance.

At lower temperatures, particularly for operation below the acid or water dew point temperature, corrosion risks have to be fully assessed in the design and construction of any precipitator in order to obtain a reasonable cost effective life span. For example, for sulphuric acid mist collection, the early plants were constructed using lead coated steel components or harder antimonial lead for discharge and collector electrodes. More recent acid mist plants have used reinforced glass fibre components for both the containment vessel and collectors, the collectors having some form of conductive coating to discharge the deposited particles, while titanium or high nickel alloys have been used for the discharge electrodes. On other plants subject to corrosion attack, earlier designs used acid resistant brick lined reinforced concrete for the casings, with wetted wooden collectors and titanium or high nickel alloys for the discharge electrodes. For the cleaning of blast furnace and coke oven gases basically devoid of oxygen, normal but thicker carbon steel components have given an acceptable life span usually greater than 20 years.

With respect to designing for the operational temperature, one is seeking to choose a material that will produce a satisfactory life at an economic cost. With wet electrostatic precipitators (WESPs), which employ water/liquor washing to remove the precipitated deposits, one must also be aware of sulphur reducing bacteria (SRBs), which may be present in the recycle liquor and unless checked by a suitable biocide will attack the iron molecules in any steel work resulting in an accelerated corrosion rate.

The gas volume being treated by the precipitator is temperature dependent according to Charles's law, and since the performance of a precipitator is related to the contact time, any increase above the design temperature will increase the gas volume and therefore reduce the overall collection efficiency.

Another impact of temperature that affects the electrical operation and hence performance of the precipitator, as previously indicated, is that as the temperature increases, the kinetic energy of the gas molecules rises, which increases the corona current flow but at the expense of a reducing breakdown voltage. These effects are illustrated in Figure 3.2, which must be considered when designing the discharge electrode system for precipitators; particularly those used in high temperature applications.

For a number of applications it is necessary to reduce the temperature by moisture injection into the waste gas stream in order to reduce the fly ash electrical resistivity to a level where reverse ionisation effects are mitigated. The effect of the moisture content of the waste gas on electrical resitivity can be appreciated with reference to Figure 3.3, which, while specifically relating to a coal fly ash, indicates that increasing the moisture content of the gas from

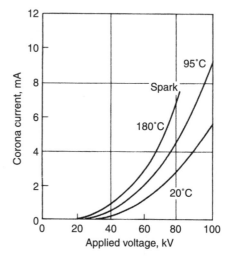

Figure 3.2   Effect of gas temperature on corona generation

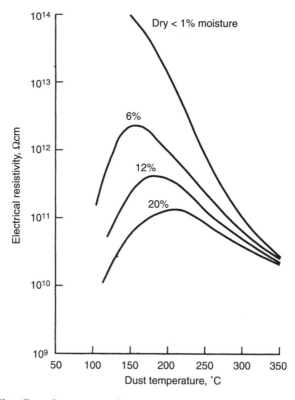

Figure 3.3   The effect of moisture in the gas on particle resistivity

5 per cent to 20 per cent reduces the electrical resistivity by two orders of magnitude.

In view of the influence of the temperature upon particle electrical resistivity and other associated temperature effects, which impact on the precipitator sizing and performance, it is imperative that the operational temperature range is fully and carefully considered for any application.

The effect of gas temperature and the electrical resistivity of the particulates upon the electrical operating conditions will be examined in much greater detail later in the chapter.

3.3 Influence of gas pressure

In general terms, the pressure of the entraining gas dictates the design and operation of the precipitator. For high pressure applications the casing can be cylindrical or ovoid in cross section, not only to contain the gases but also to withstand the operational pressure without serious distortion such as to impact on the internal alignment of the electrode system. The high pressure within the precipitator casing also complicates the removal of the collected material from the plant, because normal hopper evacuation systems, without adequate pressure reducing facilities, would allow the gases to escape into the surroundings and create a hazardous condition adjacent to the plant.

The effect of pressure on electrical operating conditions is illustrated in Figure 3.4, which indicates that at high temperature, high pressure assists in significantly raising the operational breakdown voltage. This high pressure, high

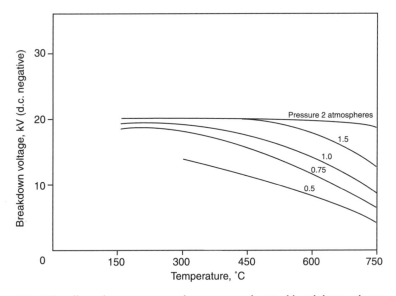

Figure 3.4   The effect of temperature and pressure on electrical breakdown voltages

temperature effect enables the electrostatic precipitator to be used for cleaning the gases ahead of gas turbines employed in integrated combined cycle gasification plants.

Converse effects arise with operation at reduced pressure, as also indicated in Figure 3.4, which impacts on the gas density such that additional corona current flows between the electrode system but at the expense of a reduction in breakdown voltage. Again the casing and dedusting system must be designed to withstand the operational negative pressure since any inleakage into the plant increases the gas volume being treated and being air, may cause the material to fire, or possibly result in an explosive gas/air mixture being formed.

3.4 Gas viscosity and density

Both the viscosity and density are related to the composition and temperature of the entraining gas. As the temperature rises, the viscosity increases, retarding the migration of the ions and charged particles; this retardation, however, is somewhat mitigated by the reduction in gas density. The effect, however, must be considered in the design stage, since the change in performance, etc., is indicated by Equation (2.14).

3.5 Impact of gas flow rate and gas velocity

It will be appreciated from Equations (2.16) and (2.17) that the sizing, and consequently the performance, of an electrostatic precipitator is dependent on the gas flow rate and hence contact time. For a given plate area A and a certain effective migration velocity ω, the efficiency is related to the gas volume V. Any change in gas flow rate effects the efficiency in order to maintain the equation balance. A typical efficiency/SCA (contact time) relationship is illustrated in Figure 3.5, for a range of typical modified migration velocities (ω_k) met in practice.

Although the above relationships hold for normal variations in gas flow rate, for dusts exhibiting poor cohesive properties, re-entrainment and scouring of the collected dust can occur by operating at too high a gas velocity, which detracts from the theoretical efficiency. At low gas velocities, two effects arise that reduce efficiency. The first is that at the lower velocity the precipitation rate (mass/unit area of collector) increases at the inlet to the field and this can upset the electrical field distribution further downstream, which significantly impacts on the overall efficiency. Secondly, for a precipitator designed for higher flow rates, the rapping frequency and intensity will not necessarily be correct for the lower velocity and hence serious re-entrainment could occur because of the redistribution of the deposited dust masses.

The curve in Figure 3.6 indicates how the theoretical efficiency based on a constant modified migration velocity (ω_k), changes with gas velocity through

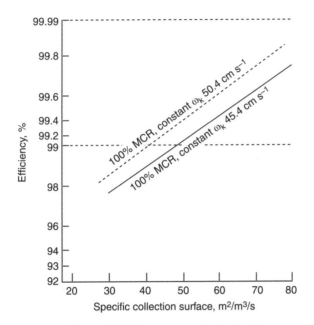

Figure 3.5   Precipitator efficiency/SCA (contact time) relationship

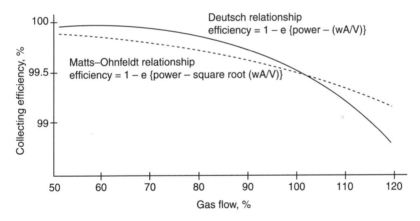

Figure 3.6   Effect of gas velocity upon precipitator efficiency

a precipitator designed with a nominal operating velocity of 1.4 m s^{-1}. At low, but decreasing velocities, the collection efficiency tends to rise to a maximum, but at higher velocities, dependent on the cohesive properties of the dust, re-entrainment results in a significant degrading of efficiency. In general, for a dry precipitator design gas velocities approaching 2 m s^{-1} should be avoided, although for wet precipitators higher velocities can be tolerated, because of the lower scouring risk.

Although dry precipitators can be designed for operation at low velocity, economics and space limitations in practice automatically tend to limit the design velocity to some 1 m s^{-1}.

3.6 Gas turbulence

Turbulence is a feature of all operational precipitators, initially arising at the inlet distributors controlling the gas flow, then from the components and structural members inside the casing. The discharge electrodes themselves produce eddies and vortices and the flow across the collector boundary layer itself creates turbulence whether or not there are protuberances on the collector surface. In addition, the electrons and negative ions moving at high velocity through the inter-electrode space, the so termed 'ionic wind', creates intense local turbulent vortices.

The turbulent gas flow is influenced by gravity, the electric field and viscosity that serves to control the flow of electrons, ions and the particles. In his early work, Deutsch [1], assuming that the forces acting on a spherical particle to be electrical Coulomb and Stokes' drag forces in a quiescent gas, proposed that the particle migration velocity, ω, took the form of

$$\omega = qE_p/3\pi d_p\eta, \tag{3.1}$$

where q is the particle charge, E_p is the field strength, d_p is the particle diameter and η is the gas viscosity.

The diffusion charge and field effects were later considered by Cochet [2] and by Smith and McDonald [3], by introducing the term 'particle mobility' (m_p), where

$$m_p = q/3\pi d_p\eta. \tag{3.2}$$

Substituting in Equation (3.1) we obtain

$$\omega = m_p \times E_p.$$

This is the Coulomb Stokes' value, which Deutsch introduced into his efficiency equation as

$$\text{efficiency} = 1 - \exp\left[(\omega/v) \times (L/d)\right], \tag{3.3}$$

where v is the mean axial velocity, L is the precipitator length, d is the distance between discharge and collector electrodes, and ω is the so termed effective particle migration velocity.

In this equation, Deutsch assumed that the particle concentration profile across the inter-electrode area was uniform owing to electric wind and that the flow was fully turbulent.

The degree of turbulence has a negative effect on efficiency particularly at low flows as indicated previously. Work by Thomsen *et al.* [4], working with parallel plate precipitators, produced curves shown in Figure 3.7, relating the turbulence intensity V_{rms}/V_0 against mean axial velocity with and without electrode

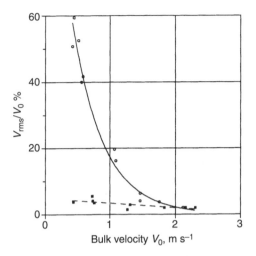

Figure 3.7   Impact of gas turbulence and velocity of performance

energisation. This graph indicates the negative impact of designing with too low an axial gas velocity, i.e. the resultant extremely high turbulence intensity when energised.

Dalmon and Lowe [5], using a large pilot scale precipitator, investigated the effect of axial gas velocity by measuring the collection efficiency for a range of particle sizes. The results of their investigation are shown in Figure 3.8. These curves indicate that the efficiency, in terms of the Deutsch migration velocity ω, increases to a point where re-entrainment and scouring of collected material causes a sharp fall in performance. Depending on the physical and chemical

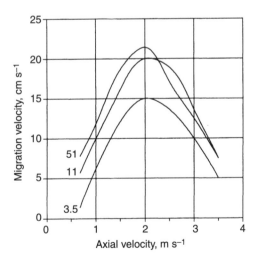

Figure 3.8   Effect of gas velocity on precipitator performance (EMV) for differing particle sizes of 3.5, 11 and 51 micro diameter

characteristics of the particles in terms of size mass, cohesivity, etc., there is an optimum design axial velocity; too low or too high a velocity would result in performance being compromised.

Underperformance as a result of operating with a gas velocity, that is too high coupled with poor cohesivity, is probably best illustrated for an electrostatic precipitator used in the power generation industry. Developments to limit NO_x emissions within the furnace area, has led to the introduction of low NO_x burners, which, although minimising NO_x formation, can produce a high carryover of large partially burnt out coal particles. These particles have a large surface area, low mass and low electrical resistivity, which under high velocity gas conditions are susceptible to re-entrainment and/or scouring and can pass through the precipitation field without being collected. In the worst case a unit with a loss of ignition (carbon carryover) of some 12 per cent at the precipitator inlet, resulted in an emission comprising 90 per cent combustible material, that is, some nine times greater than the actual fly ash.

3.7 The importance of gas distribution

On most process plants, in order to minimise cost, pressure loss, erosion, particle deposition and thermal losses through interconnecting ducts or flues, the carrier gas velocity, at full normal plant output, is typically in the order of 13–20 m s^{-1}. In order to reduce this duct velocity to around 1.5 m s^{-1} in the precipitation field, some type of diffuser is required having an expansion ratio in the order of 10 or more dependent on the design precipitator gas velocity and type of particle to be collected. The design of this transition section, where kinetic energy is converted to potential energy, usually as a pressure drop, is critical to achieve a uniform flow profile through the precipitation field if optimum performance is to be obtained.

With many installations, cost and space restrictions mean that sharp angled diffusers are used to decelerate the gas velocity and distribute the flow through the precipitator. There are two main types of diffuser used in practice; one is to use a fully splittered transition section, based on an 'egg-box' principle to spread the gas followed by a smoothing screen further downstream across the face of the field. This approach is illustrated in Figure 3.9.

The other type of approach is to use a series of perforated screens, a technique developed from wind tunnel diffusers using grids or nets. If the transition is short, having steep angles, the flow will separate from the duct walls creating a central high velocity jet, which hits the perforated plate and spreads sideways, resulting in a recirculating gas flow adjacent to the walls. To minimise this effect it is normal to have a series of perforated plates having different porosities within the transition section to obtain the degree of flow smoothing required. Although both approaches can be found in practice, the advantage of the splittered transition is a lower pressure loss of some 0.5 kPa on the overall system.

For either system to be fully effective, in terms of ultimate performance of the precipitator, it is important that the gas upstream of the transition is uniform in

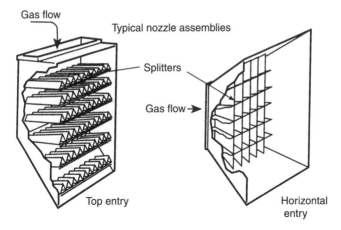

Figure 3.9   Inlet splittering to obtain uniform gas flow

terms of flow, temperature and dust distribution, any disturbance in the flow leading to non-uniformity of velocity within the precipitation field.

Although ideally, the gas velocity profile through the precipitation field should be uniform, in practice, because of boundary layer drag and areas above and beneath the collector system where the gas can expand, it has been generally agreed throughout the precipitation industry that an acceptable distribution is one where the 'root mean square' deviation is no more than 15 per cent [6].

Although most precipitator suppliers have adopted this standard, there has been a recent trend to consider a 'skewed' distribution [7], particularly in the vertical plane, for units having a high carbon carryover or very high particulate loadings. In these instances, some precipitators have had the gas distribution modified such as to increase the gas velocity in the lower part of the inlet field to be some 30 per cent higher than the top of the field, while in the outlet field the converse flow profile is adopted, as indicated in Figure 3.10.

The theory of this form of distribution is that at the inlet, with higher mass loadings or readily entrainable dusts, the bulk of the collected material has the least distance to fall after release from the collectors, which should reduce re-entrainment into the downstream sections of the precipitator. At the outlet, by having the high flow towards the roof of the precipitator, any dust released from the collectors will pass through a lower axial velocity area, which again should reduce any incipient re-entrainment.

Regardless of whether a standard or skewed approach is considered, it is important that the gas distribution in the horizontal plane is maintained as uniform as possible, otherwise the performance will be adversely affected. A further practical consideration is the avoidance/minimisation of gas looping (bypassing) over or under the actual precipitation field. This bypass results from the slightly higher pressure drop across the field because of the internals. To minimise the risk of this bypassing impacting on overall performance, small 'kicker' baffles are installed at the top and bottom of the field to break the gas

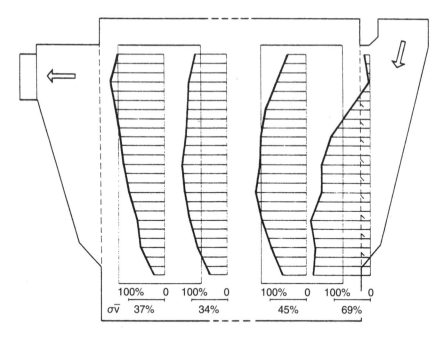

Figure 3.10   Skewed distribution profile

from the walls and divert the flow into the field area. Although this is a palliative solution, since gas will still tend to loop, plant performance measurements have produced emissions of less than 1 mg Nm⁻³ on some high efficiency installations.

To achieve an acceptable gas distribution it is normal practice to carry out testing, either in the field to confirm results obtained from large scale models of the system (~1/10 full scale) or data based on computational fluid dynamics (CFD) approaches. Although correcting gas distribution in the field is possible, time constraints and the fact that a single splitter plate on large installations can weigh some 500 kg, make this approach impracticable except for the smallest installations.

3.7.1 Correction by model testing

In the case of large scale modelling, it is important that geometrical, kinematic and dynamic similarity is achieved between the model and full scale plant. The physics of flow is described by three Navier–Stokes equations, one per coordinate in space and the equation of continuity. By making the equations dimensionless, in terms of mass, length and time, it can be shown that the solution is universal, providing the numbers appearing in the equations are equal in the model and full scale. These include the Reynolds number, *Re*, the Mach number *Ma* and the Euler number *Eu*. (Note that the Euler number is automatically fulfilled as it depends on the Reynolds number.)

The Reynolds number, *Re*, is the ratio between the flow inertia and friction

$$Re = vd/\eta. \tag{3.4}$$

The Mach number, *Ma*, is the ratio between the gas flow velocity and the speed of sound

$$Ma = v/a = v/(kRT)^{\frac{1}{2}}, \tag{3.5}$$

where *v* is the gas velocity (in m s^{-1}), *d* is the characteristic length (in m), η is the kinematic viscosity, *a* is the speed of sound (in m s^{-1}), *k* is the adiabatic constant, *R* is the universal gas constant and *T* is the absolute temperature.

When using a 1/10 scale model of a power station cold side precipitator, under the same temperature and gas conditions, 'exact similarity' would involve running the model with a field velocity some ten times higher than the full scale plant. In practice, because the gas composition, and hence viscosity and temperature, are very different, plus reducing the number of collector plates in the model effectively to increase the Reynolds number, the velocity in the model will reduce to around six times the actual for similarity. This velocity, however, may be too high in some duct configurations since the gas will be approaching compressibility, which will impact on the flow characteristics. To investigate this, by operating the model at several different but lower velocities, the significance of operating at lower Reynolds numbers on the precipitator gas distribution can be established.

Generally by operating at a model velocity of around twice that of the full scale plant, acceptable correlation with the field measurements can be obtained. During the construction of both model and full scale units it is imperative that the positioning of all splitters, baffles and any 'kicker' plates to minimise bypass are faithfully reproduced if correlation is to be achieved.

While the inlet transition determines the distribution profile through the precipitator inlet fields, a steep angled outlet transition can impact on the flow in the outlet field and it is normal for splitter plates to be included at the outlet to maintain control of the gas flow.

3.7.2 Computational fluid dynamic approach

In recent years the development of powerful and fast computers has meant that solving the differential equations describing fluid motion is now possible. Physically the conservation of mass and Newton's second law are applied to the fluid, expressed mathematically as the equation of continuity plus the Navier–Stokes equation, in two or three dimensions. Solving these equations is a discipline defined as direct numerical simulation, or generally computational fluid dynamics.

Even with today's fastest supercomputers, it would be impossible to divide the calculation domain into small enough parts to describe all the details in the flow field. Solving the equations directly is a discipline called direct numerical simulation, which is still restricted to very limited Reynolds numbers and small geometries.

With the smallest eddies, for example as in the Kolmogorov length scale $\left(v^3 \times \dfrac{\delta}{u^3}\right)^{0.25}$, where v is the kinematic viscosity $\approx 40 \times 10^{-6}$ m^2 s^{-1}, δ is a typical shear layer thickness $\approx 10^{-1}$ m and U is the bulk velocity ≈ 1 m s^{-1}, the length scale is of order 0.3 mm. This means that a precipitator for a power plant installation having approximate dimensions of $15 \times 15 \times 20$ m^3 should have a mesh number of around one hundred million millions, far beyond the capacity of most computers to tackle direct numerical simulation.

Instead, the turbulent variables are taken as average values plus fluctuating parts; for example, $u = U + u'$, where u is the instantaneous x velocity, U is the time average velocity and u' the time-dependent fluctuation (time average of $u' = 0$).

Introducing these variables into the Navier–Stokes equation and time averaging, leads to:

$$\rho \, dU_{av}/dt = \rho g - \nabla p_{av} + \nabla \tau_{ij}, \tag{3.6}$$

where the stress tensor

$$\tau_{ij} = \mu(du_i/dx_j + du_j/dx_i) - \rho(u_i' \, u_j')_{av}. \tag{3.7}$$

The first term is the laminar stress, and the second is the turbulent stress. The turbulent part of the stress tensor is either found by solving the so-called Reynolds stress equations or simply by expressing it using average flow values and the so-called Boussinesq assumption

$$\tau_i = \mu_i/dU/dy,$$

where μ_i is the eddy (or turbulent) viscosity, U is an average velocity and y is a coordinate perpendicular to the vector U.

The value of μ_i is determined by using either Prandtl's mixing length theory, that is $u_t \equiv \rho l^2 |dU/dy|$ (where l is the so-called mixing length), or by using the 'k–ε' model: $\mu_t \equiv C_\mu k^2/\varepsilon$, where C_μ is a constant or coefficient, k is the turbulent kinetic energy, $k = \frac{1}{2}(u'^2 + v'^2 + w'^2)$, and ε is the turbulent dissipation, $\varepsilon = k^{3/2}/L_d$, where L_d is a length scale for the dissipating eddies.

The more complex the turbulence model is, the more differential equations must be introduced, which can only be solved by closing the system by introducing algebraic expressions based upon empiricism.

The most common methods today are based on transformation of the differential equations for conservation of energy, mass and momentum to difference equations, which can be solved by integration after applying the given boundary conditions. Originally, in two dimensions, the equations were transformed and solved using stream-function values Ψ ($\rho u = d\Psi/dy$, $\rho v = d\Psi/dx$) and vorticity w ($w_z = dv/dx - du/dx$), thus eliminating pressure as a variable. Later, with more efficient computers, the equations were formulated and solved in the primitive variables, pressure and velocity, in three dimensions, thus avoiding the problem of defining the exact boundary conditions for vorticity having steep gradients close to the walls. As velocity gradients are also steep at the walls, special attention must be paid to the description of analytical near-wall variations, the so-called wall functions.

The difference equations are often solved in Cartesian or orthogonal grids, equidistant or almost equidistant, using 'upwind differences' and taking into consideration the local direction of flow. One problem with the grids is the appearance of 'numerical diffusion', which is a maximum if the flow vector has a 45° angle with the grid axes. In recent years, advanced grids have been developed, e.g. 'adaptive grids', where the mesh size changes according to the gradients of the variables, decreasing in size where gradients are steep. Another approach is the use of 'multigrids', where calculation shifts between a fine and a coarse mesh, thus smoothing out short wave and long wave variations.

The difference equations are normally solved indirectly (integrated) by iteration, sometimes several hundreds of iterations, until numerical stability is achieved by successive under or over relaxation approaches. The normal criterion for having found the final solution is one of mass continuity, and the calculation stops when the maximum residue is less than, say, 10^{-4}. The 'solver procedure', being a chapter in itself outside the scope of this book, typically uses various numerical principles, continually updating the values in the domain as soon as new values have been found. A number of fast and efficient solvers are now commercially available and are in common usage for CFD work.

With more complicated geometries, such as precipitators including complicated transition pieces, guide vanes and screens, grid designing is difficult. A simple method is to use a parallelepipedal domain and block out all the outer elements until an approximate contour is achieved, leaving the geometric domain boundary as a step surface. This means that there is a limit to how precise the solution close to the walls can be, even though the internal flow is only slightly influenced. In fact, it is a grid generator, which is the main issue for the operator. The easier the grid generation, the more different configurations can be calculated in a reasonable time, presuming, of course, that the solver is effective and fast. Less than 2 h per new contour and less than 1 h per modification would be considered acceptable.

The real progress in mesh generation is a mesh that can be fitted without any consideration of the interfacing, at least as seen from the point of view of the user. Up to now, mesh structures have had to fit where ducts and transition pieces, or transition pieces and precipitator housing, meet, but new improved grid systems are making it possible to have an essentially polar mesh of a circular configuration, corresponding with a rectangular mesh of a box, without any concern about mesh fitting at the plane where they meet. This block type of approach is well suited for duct and precipitator analysis with respect to both gas distribution and pressure drop determination. Figure 3.11 shows this type of mesh from the code Star-CD [8]. Other forms of program are commercially available from, for example, Fluent Inc. [9].

The more advanced the mesh, the more variables to be treated and the stricter the convergence criteria, the more calculation time is needed to find a correct solution. The need for space in the memory and on the disc is another 'eye of a

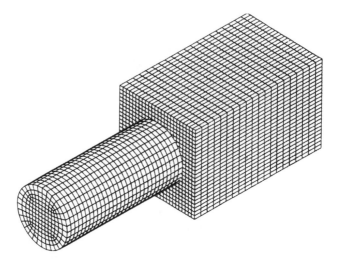

Figure 3.11   Typical solver mesh from the code Star-CD

needle', the cell number increasing roughly with the product of the number of variables and the reciprocal of the mesh size to the third power. While details such as guide vanes, kicker plates and ladder vanes have to be modelled separately, there is a possibility to simulate perforated screens using some sort of porosity model such as a Darcy porosity.

Screens with evenly distributed guide vanes might be modelled as a whole, but this calls for special routines, not normally commercially available. In the equations the so-called 'source' terms can be modified in order to reflect the influence of the gas distribution screens [10]. Screens can also be modelled by 'blocking' out cells, but this procedure demands a very fine mesh with many nodes. The flow before a perforated screen is often recirculating, unless small screen guide vanes are used. Such flows can be identified using CFD, revealing velocity vectors parallel to the screen, making it impossible to improve the distribution by modifying the screen by increasing or decreasing the open area alone. Programs that cannot treat small thin oblique surfaces in an effective and easy way should not be used. As in the use of porosity for simulating a perforated screen, it is possible to design a subroutine for a screen combined with distributed guide vanes and use it as a black box when the calculation is operated in the screen domain, which would reduce the need for a complex mesh with a large number of grid points.

3.8 Effect of particle agglomeration

Particle agglomeration is attributable in the main to the collision and impaction of the smaller sized particles, which are in continuous motion exhibiting what is termed Brownian motion, as the result of bombardment of the particles by gas molecules. The effect of the bombardment is maximised for the finer particles and, as they agglomerate, Brownian motion decreases so the possibility of further collision is reduced.

In precipitator applications, the effect of fine particle agglomeration is to shift the particle sizing upwards; this will impact on their free falling velocity according to the Stokes–Cunningham relationship and may reduce the possibility of space charge and corona suppression effects.

Another aspect, which is gaining prominence in the air pollution field, is that of the need efficiently to collect toxic and heavy metals. In most processes these materials are very small in terms of mass concentration, but recent toxicology investigations have indicated their importance to health. Heavy metals usually exist in the downstream area of the process as condensed submicrometre particles after passing through a volatilisation stage in their life cycle. During condensation, some of the particles use the larger particles as condensation nuclei, so it is not unusual to find small metal rich fume adhering to larger inert particles, as indicated in Figure 3.12. This means that these fine heavy metal particles are removed along with the easier to collect coarser particles. This would help to explain why solid phase toxic and heavy metal mass balance measurements, in spite of their smaller particle size and potentially lower

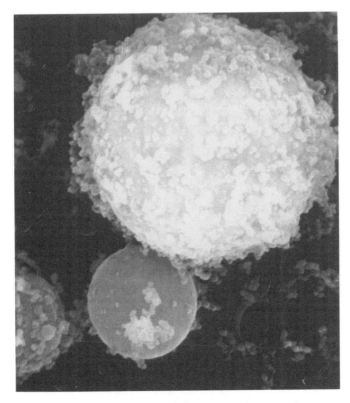

Figure 3.12   Photograph of a large particle with adhering finer particles

performance, compared with the coarser particles, produce almost the same order of efficiency as the bulk materials [11].

The effects of agglomeration and cohesion, although producing particle floc- culation, are the result of different mechanisms and should not be confused. Agglomeration, as stated above, arises wholly in the gas phase because of Brownian motion and, except for impaction on some larger particles, predomin- antly applies to small submicrometre particles forming larger but still small units (see Figure 3.13 [12]). Cohesion, on the other hand, applies to the collected precipitated dust, as will be covered in Section 3.9. Cohesion is a measure of the binding mechanisms holding together all size particles, as a result of electrical and mechanical forces acting on the layer rather than individual particles.

One of the heavy metals that is causing concern, particularly in the USA, where legislation is scheduled to be enacted by December 2007, is the present uncontrolled emission of mercury from power generation plant because of the widely reported potential health problems. Measurements have indicated that the annual total emission of mercury from US power plants is in the order of 50 t. Mercury, unlike most other heavy metals, because of its low melting point, −40 °C, can exist in a solid, a liquid or gaseous phase, depending on the chemical form and the final gas temperature, which complicates its removal.

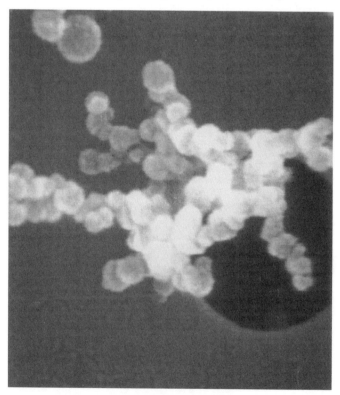

Figure 3.13   Photograph of agglomerated fine particles

The effect of gas temperature on vapour phase concentration of a number of

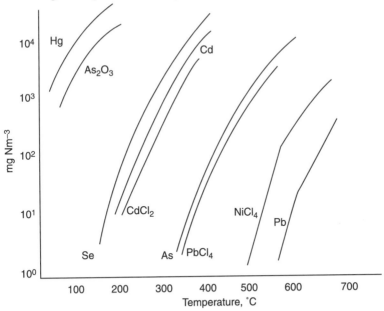

Figure 3.14   Effect of temperature on vapour phase concentration of various heavy metals

different compounds is indicated in Figure 3.14 [13]. This shows that additional material can be condensed from the gaseous phase by reducing the operating gas temperature, such that an enhanced overall removal efficiency can be obtained by the precipitator.

If it should be necessary for the gas temperature to be reduced to approach the water dew point, then a wet, as opposed to dry, precipitator can be used to attain the higher removal efficiency. Although the cost of a water treatment plant associated with a wet precipitator installation can be expensive, if the installation follows a wet desulphurisation plant, then a separate water treatment plant is not required, since the flushing liquor can be the feed water to the scrubber and the effluent fed back into the scrubber water treatment system [14].

This type of approach is currently being used to clean the gases exiting wet desulphurisation plants applied to a number of power plants. Although the wet scrubber is effective in lowering the sulphur dioxide levels to meet current legislation and in further reducing the particulate emission from the upstream precipitators by approximately 50 per cent, the very act of reducing the gas temperature to the water dew point temperature automatically results in the condensation of any upstream gaseous phase components, such as free sulphur trioxide, which forms an acid mist. The low pressure drop across the scrubber is too small to collect the fume sized acid mist and although reheat is used to assist in plume dispersion, this is too low a temperature to evaporate the acid, which can produce an undesirable and visible plume. A further complication is that nucleation can also arise between the acid mist and other compounds in the exit gases further adding to the discharge. To mitigate this emission, wet type precipitators are being included in the system to polish the gases such as to improve the environmental aspects of the discharge. Some installations have standalone electrostatic precipitators (ESPs) following the scrubber, while other installations have the wet ESP mounted in the scrubber tower itself [15], either approach being capable of removing the acid mist particulates.

While the above approaches can readily collect the liquid and solid phase materials, if gaseous phase removal is required, then an absorption technique can be applied. In this instance, activated carbon for example, or some other suitable form of 'getter', can be injected upstream of the precipitator at a point which gives a contact time of around 1 s, such that absorption of the heavy metals onto the activated carbon can occur. The activated carbon contaminated with the mercury and or other heavy metal gaseous phase components can then be removed from the gas stream by a conventional dry precipitator.

3.9 Particulate cohesivity effects

In electrostatic precipitation, particle cohesion plays an important role in plant performance, firstly in respect of how the particles are held to the collector and secondly how the particles hold together in their transference to the hopper, following rapping.

After reaching the collector plate, particles are held initially by electric forces and after losing their charge would, without the mechanical binding force, be readily released and possibly re-entrained by the gas stream. From Stokes law, with a precipitator gas velocity of 1.5 m s^{-1}, only bound particles greater than 500 μm equivalent diameter stand any chance of reaching the hopper. Without this particle cohesion the dust would be subject to massive re-entrainment, which would have a deleterious impact on efficiency.

The forces holding the particles to the collectors and themselves are a combination of electrical and mechanical van der Waals forces. Both are linked to the surface chemistry of the particle, the nature of which depends not only on the composition of the dust but also on the gas constituents to which the particle has been exposed. For particles having electrical resistivities above 10^{11} Ωcm, the electrical holding force tends to predominate over the mechanical force, but for low resistivity dusts the mechanical force becomes more dominant, depending on the dust layer properties.

In general, light fluffy dusts adhere poorly, while dense or sticky dusts adhere well. Particles such as carbon and ionic salts, which can form 'snow flakes', have low packing densities and poor cohesion, so the particles are only loosely held together. An exception is where the ionic salt is deliquescent/hygroscopic and absorbs moisture from the flue gas, which cements or binds the particles together. If the temperature should approach the gas dew point the dust becomes sticky and may be difficult to remove from the internal components of the precipitator.

The importance of cohesivity and precipitator performance has been discussed in general terms by a number of investigators; for example, Lowe and Lucas [16] calculated the forces to which a dust particle would be subjected on the collector; Penney and Klinger [17] measured the cohesion of an electrostatically precipitated dust layer after the power had been switched off and compared the result with that of a mechanically formed layer with fairly good agreement.

Dalmon and Tidy [18] determined the relative importance of cohesion for both the mechanical and electrical forces and then related these cohesive properties to precipitator performance. This investigation was carried out with two very different power station fly ashes, one relatively fine having high resistivity and another coarse containing 33 per cent carbon and of low resistivity. Both dusts in practice gave rise to poor precipitation efficiencies. The collected dust samples were initially washed to remove all surface conditioning effects/agents and injected into an oil fired combustion rig fitted with a 250 mm hexagonal tube precipitator. Precipitator efficiency measurements were made with the dust being self-conditioned by the flue gas and then with various conditioning agents known to reduce resistivity being injected into the rig upstream of the precipitator with about 0.5 s exposure time.

At the end of each run, a detachable section of the collector was removed and the bulk density and the minimum compressive load, i.e. the load to just observe further compression of the dust precipitated on the sheet, were determined. Dust from the collector plates at the end of each run was placed in a powder tensiometer [19] and the ultimate tensile strength (UTS) of the deposited layer was determined for each sample over a range of bulk densities. (Although the cohesivity of the powder sample was not directly measured, Farley and Valentin [20] have shown that for powder beds the UTS is directly proportional to the cohesivity.)

The results from these measurements showed that each dust had a distinct but separate relationship for UTS against compressive load, which was independent of any conditioning agent, and was basically a function of particle size, shape and hardness. The force required to fracture the bed was proportional to the product of the mean strength of the individual forces and their number density in the fracture surface. For a specific ash, the tensile strength at a fixed bulk density can be theoretically used as a measure of the particle/particle force, or cohesion, and thus can indicate the relative ease of re-entrainment.

The measured precipitator performance, when treating variously conditioned high resistivity fly ashes, showed little dependence on the UTS value measured at the precipitated bulk density or the UTS at a constant bulk density. There was, however, as expected, a distinct effect of resistivity on collection efficiency owing to the conditioning agents and hence the conclusion, for high resistivity dusts, is that the improvement in performance is primarily by reduction in electrical resistivity, rather than reduction in the re-entrainment value.

With coarser low resistivity ash (10^8 Ωcm), conditioning agents were not expected to produce any further reduction in resistivity; however, their addition to the flue gases was found significantly to improve the precipitator perform-ance. Collection efficiencies for the fly ash and carbon were separately evaluated for a range of injection rates of SO_3 and ammonium sulphate; these are reproduced in Figure 3.15.

Figure 3.15 clearly demonstrates that it is the carbon efficiency that is improved by the conditioning agent by preventing re-entrainment. When the UTS was measured at constant bulk density, the relationship of precipitation performance (emv) against UTS was found to be linear (Figure 3.16), with the higher performance emvs being associated with the higher UTS. This increase in the individual particle/particle cohesive force is responsible for the improved collection efficiency of the carbon and reduced re-entrainment potential.

Dalmon and Tidy [18] concluded that tensile strength measurements made on mechanically formed beds of highly resistive dusts gave results that closely align to those measured on a precipitated bed after the field was removed. Cohesivity will be present in any deposited layer and will be augmented by an equal force, due to the electrical field, when the ash resistivity is high. As the ash resistivity reduces, the electric field effect reduces and may even become repulsive, so par-ticles are readily re-entrainable. Increasing the tensile strength of low resistivity particles by the injection of conditioning agents produced increased precipita-tion efficiency, mainly from the retention of the very low resistivity carbon particles, as a direct result of the higher prevailing cohesive forces.

The effects of agglomeration and cohesion, although producing particle flocculation, are the result of different mechanisms and should not be confused. Agglomeration, as stated above, arises wholly in the gas phase because of Brownian motion and, except for impaction on some larger particles,

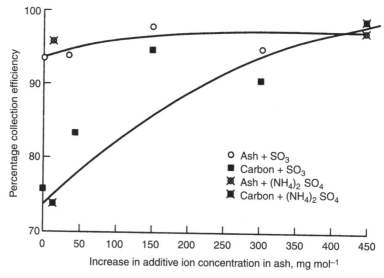

Figure 3.15   Individual fly ash and carbon efficiency relationships with conditioning

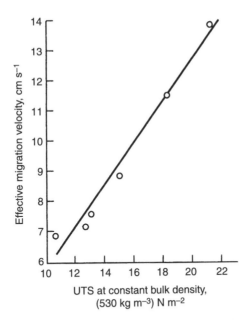

Figure 3.16   Effect of the UTS of the deposit on precipitation performace (emv)

predominantly applies to small submicrometre particles forming larger but still small units. Cohesion, on the other hand, applies to the collected precipitated dust. Cohesion is a measure of the binding mechanisms holding together all size particles, as a result of electrical and mechanical forces acting on the layer rather than individual particles.

3.10 References

1 DEUTSCH, W.: 'Bewegung und Ladung der Elektrizitätsträger im Zylinderkondensator', *Ann. Phys.*, 1922, **68**, pp. 335–44
2 COCHET, R.: 'Lois Charge des Fines Particules (submicroniques) Etudes Théoriques Controles Récents Spectre de Particules'. Coll. Int. la Physique des Forces Electrostatiques et Leurs Application', Centre National de la Recherche Scientifique, Paris, 1961, **102**, pp. 331–8
3 SMITH, W. B., and McDONALD, J. R.: 'Development of Theory for the Charging of Particles by Unipolar Ions', *Journal of Aerosol Science, 1976,* **7**, pp. 151–66
4 THOMSEN, H. P., *et al.*: 'Velocity and Turbulence Fields in a Negative Corona Wire-Plate Precipitator', *Proceedings 4th Symposium on the Transfer and Utilisation of Particulate Control Technology*, Houston, Tx., USA, 1982, II EPA, 600/9-025b, 1984
5 DALMON, J., and LOWE, H. J.: 'Experimental Investigations into the Performance of Electrostatic Precipitators for P.F. Power Stations'. Coll.

Int. la Physique des Forces Electrostatiques et Leurs Application, Centre National de la Recherche Scientifique, Paris, 1961, p. 102

6   ICAC Publication No. EP 7: 'Gas Flow Model Studies', Institute of Clean Air Companies, 1993

7   HEIN, A. G., and GIBSON, D.: 'Skewed Gas Flow Technology Improves Precipitator Performance – Eskom Experience in South Africa'. Proceeding 6th ICESP Conference, Budapest, Hungary, 1996, pp. 238–43

8   STAR-CD: Version 3.1 User Manual Computational Dynamics Ltd., London, 2000

9   FLUENT INC.: Fluent 4.4 Users Guide, Lebanon, New Hampshire, USA, 1996

10 NIELSEN, N. F., *et al.*: 'Numerical Modelling of Gas Distribution in Electrostatic Precipitators'. Proceedings 8th ICESP Conference, Birmingham, Al., USA, 2001, Paper A2–2

11 PARKER, K. R., and NOVARGORATZ, D. M.: 'Electrostatic Control of Air Toxics'. Proceedings EPRI/EPA 9th Particulate Control Symposium, Williamsburg, USA, 1991, Session TR 100471. 2

12 KAUPPINEN, E. I., *et al.*: 'Fly Ash Formation on PC Boilers Firing South African and Colombian Coals'. Proceedings EPRI/EPA International Conference on *Managing Hazardous and Particulate Air Pollutants*, Toronto, Canada, 1995

13 SANYAL, A., and PARKER, K. R.: 'Challenges to Control Mercury and Other Heavy Metal Materials'. Proceedings Power-Gen 2001, Las Vegas, Na., USA, 2001

14 PARKER, K. R.: 'WESPS and Fine Particle Collection'. Proceedings 8th ICESP Conference, Birmingham, Al., USA, 2001, Paper B5-1

15 KUMAR, S., and ELSNER, R. W.: 'Analysis of Wet ESP Performance at Xcel Energy's Sherbourne County Generating Station'. Proceedings 8th ICESP Conference, Birmingham, Al., USA, 2001, Paper B5-3

16 LOWE, H. J,. and LUCAS, D. H.: 'The Physics of Electrostatic Precipitation', *Brit. Jnl. Applied Physics*, 1953, Supplement No 2

17 PENNEY, G. W., and KLINGER, E. H.: 'Contact Potentials and the Adhesion of Dust', *Trans AIEE*, 1962, **81** (1) pp. 200–4

18 DALMON, J., and TIDY, D.: 'The Cohesive Properties of Fly Ash in Electrostatic Precipitation'. Atmospheric Environment (Pergamon Press, London, 1972) **6**, pp. 81–92

19 ASHTON, M. D., *et al.*: 'An Improved Apparatus for Measuring the Tensile Strength of Powders', *Journal of Scientific Instruments*, 1964, **41**, pp. 763–5

20 FARLEY, R., and VALENTIN, F. H. H.: 'Effect of Particle Size upon the Strength of Powders', *Powder Technology*, 1968, **1**, pp. 344–54

Chapter 4

Mechanical features impacting on electrical operation

This chapter will consider the practical aspects of the previously presented theory and how this impacts on the actual design and operation of an electrostatic precipitator.

4.1 Production of ions

From the foregoing discussion it will be appreciated that in order successfully to precipitate particulates, the electrostatic precipitator requires both corona current flow, in the form of ions to charge the particles, and an electric field to cause migration of the charged particles.

In both the tube and parallel plate form of precipitator, the collecting electrodes, in relation to the discharge elements, can be regarded as having an infinite radius of curvature. The much larger radius of the collecting electrodes means that they do not emit electrons and ions except in special circumstances, which are related to the formation of reverse or back corona resulting from a highly resistive dust deposit, as will be discussed later.

For all precipitator designs, there is a relationship between the voltage and corona current formation. This relationship is dependent on the form of discharge element in terms of radius of curvature, which dictates the local field intensity, together with the gas temperature, gas composition, gas pressure and whether particles are present in the inter-electrode area. At a certain voltage, the local field intensity adjacent to the element reaches a point where ionisation of the gas molecules arises and ions and electrons begin to be emitted; this is termed the ionisation voltage. As the voltage is further increased, the number of ions produced increases, resulting in a higher corona current, reaching a maximum value just at the point of electrical breakdown between the discharge element and collector.

The theoretical determination of the current–voltage characteristics for the wire–plate geometry is, however, extremely difficult, requiring solutions to the quasi-state electro-dynamic Maxwell equations, relating the electric field, the electrode potential and space charge with the resultant current density field. The major difficulties arise through the peculiarities of the electrical conduction in gases, particularly as the electric field which transports the ions also creates them and also where the ionic space charge associated with the electric field is comparable with the electrostatic charge on the surface of the electrodes. That is, the ions themselves distort the electric field that produces and transports them. These characteristics produce a strong inter-relationship between the space charge density and electric field, which means that they cannot be solved separately. Solutions, however, have been evaluated for the Maxwell equations, using analytical numerical approaches (CFD), where the current density is evaluated for various positions within the inter-electrode area for different voltages. The average current density for each location is then calculated for each voltage level and position as indicated in Figure 4.1.

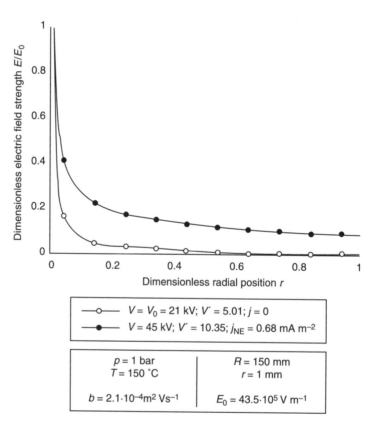

Figure 4.1   *Current density calculations at different locations within an electric field*

From the three Maxwell equations governing the electric field, Poisson's equation can be derived as

$$\nabla \times E = \rho/\varepsilon_0, \qquad (4.1)$$

where E is the field strength, ρ is the charge density (C m^{-3}) and ε_0 is the permittivity of free space (8.85×10^{-12} F m^{-1}).

Because of the difficulties of theoretically determining the voltage–current characteristics, most suppliers empirically determine the characteristics under laboratory conditions by using an air load and then applying corrections for presumed operating conditions. Some typical electrode forms found in modern electrostatic precipitation installations are illustrated in Figure 4.2 [1]. Although it is possible that these designs fall short of the theoretical 'ideal' form of

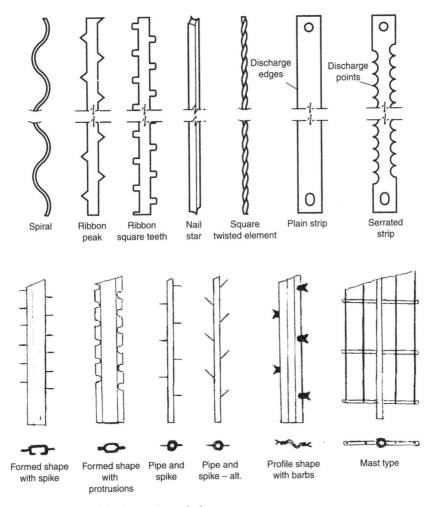

Figure 4.2   Typical discharge electrode forms

discharge element, because some 50 km of element can be found in large installations, the actual electrode form used is dictated more by the ease of fabrication, availability and other cost considerations. Because present collection
efficiencies, well in excess of 99.5 per cent, can be readily attained, it would
be difficult to justify higher costs for 'improved' designs, unless significant
performance gains would result.

4.1.1 Quasi-empirical relationships

If one considers a wire–plate configuration, as indicated in Figure 4.3b, then
simplifying the approach, by assuming that the current is small and the change
of potential by the ionic space charge is represented by the more simple wire–
tube geometry, see configuration Figure 4.3a, then for the wire–plate form, the
average current density, j_s, as a function of the element voltage (potential), can
be expressed as

$$j_s = 8\pi \times \varepsilon_0 b/c \times s\, 2 \log_e(9d/r_0) \times V(V - V_c)\ \text{A m}^{-2}, \qquad (4.2)$$

where b is the ion mobility (approximately $2.1 \times 10^{-4}\ \text{m}^2\ \text{Vs}^{-1}$ for negative corona
in air), V_c is the corona onset voltage, V is the applied voltage, d is the equivalent
cylindrical radius for the system ($d = 4\ s/\pi$ for s/c values ≤ 0.6), c is the discharge
electrode separation, s is the collector separation and r_0 the effective radius of
the electrode.

The corona onset or inception voltage V_c can be expressed by

$$V_c = r_0 \cdot E_c \cdot \log_e(d/r_0). \qquad (4.3)$$

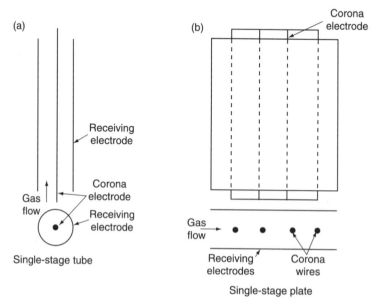

Figure 4.3   *Typical wire–tube and wire–plate precipitator formats*

The corona onset field, E_c, has been determined empirically for a large number of electrode forms and for negative energisation in air can be expressed by

$$E_c = \delta[32.2 + 8.64 \times 10^4/(r_0/\delta)^{\frac{1}{2}}] \tag{4.4}$$

where δ is the relative gas density compared to air at 1 bar pressure and 25 °C.

With the coaxial geometry of the wire–tube precipitator arrangement, although the same criteria concerning the ionic flow relating to the electric field exists, it is possible to derive semi-empirical relationships similar to the wire–plate arrangement.

The corona initiation voltage can be expressed as

$$V_c = E_c(\delta \cdot r_0) \times r_0 \times \log_e(R/r_0), \tag{4.5}$$

where R is the effective radius of the collector and r_0 the effective radius of the discharge element.

The corona initiation field strength E_c is expressed as

$$E_c = A \cdot \delta + B \cdot (\delta/r_0)^{\frac{1}{2}}, \tag{4.6}$$

where A and B are constants.

Figures 4.4 and 4.5 show typical voltage–current relationships, indicating the corona threshold voltages for various gas mixtures and for different discharge element diameters for conventional wire–tube precipitators.

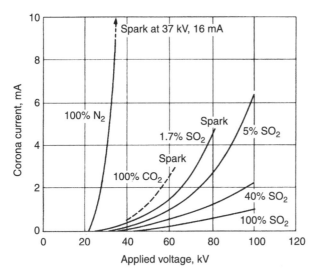

Figure 4.4 *V/I characteristics, indicating the corona threshold voltage for different gas mixtures*

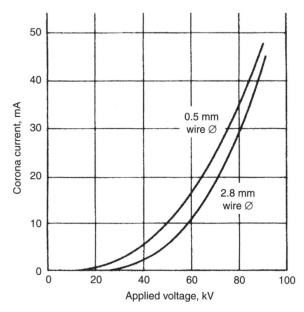

Figure 4.5   *VII characteristics indicating the corona threshold voltage for different electrode radii in air*

4.2 Discharge electrode forms

The type of discharge electrode used over the past century has taken many different forms and designs, from the earliest reported by Lodge [2], as barbed wire, and by Cottrell [3], as a pubescent electrode comprising a semi-conducting fibrous material which produced a continuous corona at a relatively low voltage of 15 kV, through to present complex designs, which are supplied and 'termed' by the suppliers as corona controlled, 'unbreakable' types [1].

It should be remembered that at the turn of the twentieth century, insulation materials were limited to naturally occurring mica, oiled paper, slate and similar materials, which severely restricted operational voltages and hence collector/discharge electrode spacing. The HT discharge electrode suspension of these early units was typically insulated from the collectors/casing by means of salt glazed earthenware tubes, rather than the later high voltage ceramic forms of insulator.

Most of the early installations were of the wire–tube design, many using a simple 2 mm diameter weighted wire as the discharging element, the weight being necessary to hold and tension the wire along the tube axis. The simple round wire of these early installations was superseded by square and hexagonal shaped wires having greater mass to withstand flashover and rapping vibration whilst having similar discharge emission characteristics.

The practical significance of the space charge effect and corona suppression became recognised when handling large quantities of fine fume and as the target efficiency demands increased. This led to the development of high emission or controlled emission electrodes, which are distinguished by having sharp protrusions where, as a result of the high field intensity at the manufactured predetermined sharp point location, optimum corona develops. Figure 4.2 above indicates a range of discharge electrode formats of the controlled emission type, typically as used by different precipitator manufacturers, together with their experimentally determined 'standard' air voltage–current characteristics as shown in Figure 4.6.

Because of the quantity of material used in the fabrication of the above electrode formats, with the exception of the spiral wire, most are classified by the suppliers as 'unbreakable' electrodes. This is partially true in comparison with the weighted wire form of electrode, which is subject to failure as a result of flashover across the electrode system, electric field induced oscillation, rapping vibration and spark erosion at the fixings. It must be appreciated that only a single electrode has to fail to produce a resultant drop in overall precipitator performance, since it impacts on the electrical operating conditions within that particular field section.

As efficiency demands, improved reliability and gas volumes have increased, to minimise cost and plant 'foot print', precipitators have become much larger and taller. With modern fabrication techniques and procedures, the present trend is to use collector plates up to 5 m wide by 15 m high, particularly on dry precipitators, which precludes the use of weighted wire forms of discharge element and some type of 'unbreakable' discharge electrode is the 'norm'. It will be appreciated from Figure 4.2 that this form of electrode can be readily mounted within a frame, which in addition to securing better alignment also assists in distributing the transmitted rapping blow throughout the discharge electrode system.

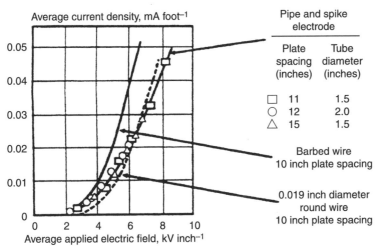

Figure 4.6   Typical voltage–current characteristics for various electrode formats

The selection of a discharge electrode form for a specific duty demands careful consideration, since each form of electrodes has its own specific emission characteristic. Although the controlled emission electrode has excellent high corona current capabilities, making them ideal for handling fine particles by helping to overcome space charge or corona suppression effects, in the absence of suppression conditions, the total corona current developed can be significantly higher than a normal lower emission type. This not only means that the transformer rectifier equipment must be much larger, but the power consumption will also be significantly higher, for the same level of efficiency. Figure 4.7 indicates a comparison of a 'standard' electrode emission against the high emission form of electrode under air conditions, which shows the higher corona electrode having an emission of some three times that of the simple wire electrode for the same operating voltage.

The effect of space charge or corona suppression can be readily observed in any precipitator installation handling a gas containing particulates. Table 4.1 indicates the electrical operating characteristics of the inlet and outlet field for a 400 mm spaced plate type precipitator application handling fine

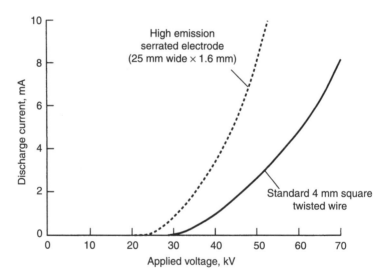

Figure 4.7   Discharge relationship between standard and high emission electrodes

Table 4.1 Operational field characteristics for a precipitator handling fly ash from Orimulsion

| | Field 1 | | Field 2 | | Field 3 | | | |
	kV	mA	kV	mA	kV	mA	kV	mA
A side	62	392	73	601	64	1014	64	1052
B side	65	663	69	517	67	1019	64	1037

fume (70 per cent being submicrometre) arising from the combustion of Orimulsion [4].

As a result of space charge, owing to the number of particles present in the inter-electrode area, the average current density for the inlet field is significantly less than that at the outlet for any given operating voltage. In practice, because of the higher ripple voltage associated with the much larger current flow in the outlet field, the average indicated voltage level appears to be reduced, although the peak voltage levels for both sections would be similar for a plant with good alignment. This apparent anomaly of an increasing current and falling mean voltage between the inlet and outlet fields is symptomatic of a correctly operating precipitator, the inlet fields also generally exhibiting a greater tendency for flashover and breakdown.

4.3 Spacing of discharge electrodes

With any design of precipitator, the separation of the discharge electrodes within the duct can significantly influence the electric field and the amount of ionisation produced. For tube type precipitators the number and location of the discharge electrodes for optimum performance is fixed, whereas with the plate type there is an optimum location for the discharge electrodes within the duct walls.

In practice and from theory it can be shown that the optimum discharge electrode spacing is approximately equal to half the duct width. Too close an electrode spacing will result in corona suppression between adjacent elements leading to reduced corona current flow and will approach a non-emitting system if the elements are too closely spaced. By spacing out the electrodes more widely in the ducting, although the corona suppression situation will be minimised, there is a point where, while maximum corona is developed for any one discharge electrode, the total specific current (in mA m^{-2}) decreases to such an extent that optimum performance is compromised.

Figure 4.8 presents an experimental plot of a simple electrode wire spacing in a 400 mm wide duct against specific current; this illustrates the effect of both too wide and too close a spacing (location) for the discharge electrodes. Although the specific current data in this figure were obtained for a fixed voltage level, even raising the voltage level for the closely spaced electrodes during the measurements did not significantly change the specific current values. From this and other measurements and investigations using simple electrode forms, the optimum spacing for maximum corona current development appears to be equal to half the duct separation irrespective of actual duct width. For more complex controlled emission electrodes, particularly those having protrusions lying parallel to the duct walls, the above half duct spacing may not be correct since emission from adjacent points may suppress each other. For these types of electrode, the optimum element spacing should be obtained either from experimental procedures or by a CFD approach [5].

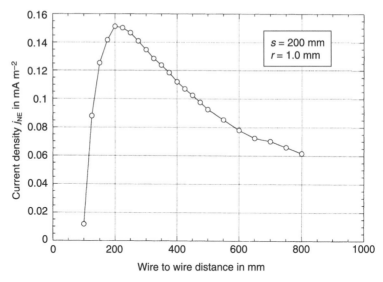

Figure 4.8   Effect of electrode spacing on specific current

4.4 Collector electrodes

It will be appreciated that for any form of precipitator the collector electrode should be 'non-emitting' as regards corona emission. This is achieved by having a large radius of curvature in comparison with the discharge elements.

In the case of the vertical flow unit, although the tube form of collector has ideal characteristics as regards maximising the electric field, it is an expensive construction because only the inner surface of the tube is used for particle deposition. To reduce costs, some suppliers have used a hexagonal approach, where flat sheets are formed into a half hexagonal shape, so that by butting two sections together they form a vertical hexagonal collector, and other suppliers have used concentric tubes of differing diameters. The advantage of these approaches is that both sides of the collector are employed for particle deposition, but the profile of the collector needs to be vertically true and without protrusions in order to minimise premature breakdown.

With horizontal flow units having collector dimensions of up to 5 m × 15 m, the collectors usually comprise a number of roll formed channels, typically fabricated from 1.6 mm thick material, mounted between heavy upper and lower members to attain the required degree of straightness and stiffness. Typical forms are indicated in Figure 4.9.

For optimum performance with the largest collectors, the degree of straightness that one is seeking is a tolerance of around ±10 mm. For designs that employ flat sheets, these are usually fairly small, e.g. 2 m × 5 m and then the edges are usually stiffened; other forms used on some wet precipitator applications are pre-stretched material of thickness 4 or 5 mm, but again the

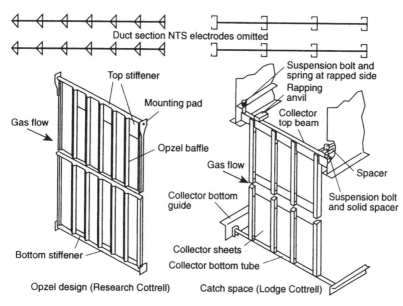

Figure 4.9 Typical collector sheet constructions

straightness and alignment tolerances required are similar, if satisfactory performance is to be obtained.

Although most dry type precipitators have metal collectors, for modern continuously irrigated wet precipitators, some designs use GRP or plastic collector plates where the liquor used for cleaning provides the necessary conductive path for discharging the particles. For wash down type units some applications employ collectors having a conductive carbon fibre coating as the conductive path.

4.5 Specific power usage

As already indicated, to achieve optimum efficiency, the precipitator requires both adequate corona generation for effective particle charging and sufficient voltage to produce the necessary field for speedily precipitating the charged particles. Since the particle migration velocity is proportional to the voltage squared (Equation (2.14), it follows for a given electrode configuration, which determines the voltage–current relationship, that the efficiency is related to the specific power input (W am^{-3}).

It can be shown that the performance, or collection efficiency, for a free running precipitator on a specific application, can be related to the total power consumed, simply as kV × mA for the gas flow in question, or more precisely the specific power usage in (kV × mA) am^{-3}. Although secondary volts and current were originally used in the evaluation, with modern instrumentation it should be possible to use either secondary or primary power metering, assuming the transformer losses are known or can be measured. Figure 4.10 illustrates such a relationship for a power generation plant precipitator installation. The significant increase in specific power required to achieve efficiencies in excess of 99.5 per cent is related to the reduction in space charge effect leading to a much higher corona generation and hence power consumption as the gas becomes cleaner, i.e. reducing particle concentration towards the outlet end of the precipitator.

On most industrial applications there is always a small percentage of fine particulate material present in the waste gases. These fine particles typically arise from the condensation of initially volatilised feed stock contaminants in the cooler reaches of the plant. As the required collection efficiency approaches 100 per cent, the effect of the reduction in the fractional efficiency with reducing particle sizing becomes significant. Therefore, to achieve the necessary efficiency a much larger plate area and hence power usage is required.

The form of the specific power/efficiency curve remains similar for all applications although the actual value of specific power will depend on the discharge electrode emission characteristics, particulate size analysis, gas composition and temperature, etc., and to a certain extent the precipitator construction details. Generally, however, it can be considered that the higher the power that a precipitator will usefully absorb, without running into arcing conditions on any field, the higher the efficiency attainable.

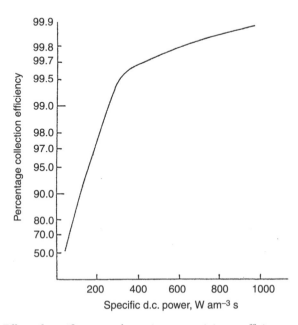

Figure 4.10   Effect of specific power absorption on precipitator efficiency

4.6 Precipitator sectionalisation

To optimise the overall performance of a precipitator treating a large gas flow rate, as on modern power plants, it is common practice to split the precipitator into separate bus sections, both in the width and length, and to energise each section with its own transformer rectifier equipment.

Splitting the unit across the width also allows any significant temperature or gas flow stratification to be more readily accommodated by having a number of electrical sets rather than just one, which would operate only under the least onerous operating condition. In addition, as an alternative to splitting the sections in width, most precipitator installations have more than one flow, so that if an outage is necessary the plant can still operate at part load. By dividing the unit in length, as the gases move towards the outlet and become increasingly dust free, the decreasing space charge effect and hence higher corona currents are more able to be controlled/optimised by their individual TR equipment.

Although current designs of rectifier equipment can have current ratings up to some 2000 mA and can safely energise some 10 000 m² of total collector plate, it has been found that as the total plate area energised by a single rectifier equipment increases it has a detrimental effect on performance. Not only does a flashover on one electrode have a larger impact on the overall efficiency, but the increase in capacitance of the larger section being energised results in a drop in the specific corona discharge, such that the total power input and hence performance decreases. This effect is indicated in Figure 4.11 covering a range of US power plant installations [6].

Figure 4.12 indicates how the specific corona current, produced from a 4 mm square section wire discharge electrode for a duct spacing of 300 mm, reduces as the size of the field energised by a single rectified set increases. The curves relate to precipitators treating normal fly ash at temperatures up to 150 °C and

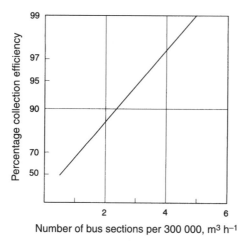

Figure 4.11   Effect of precipitator sectionalisation on performance

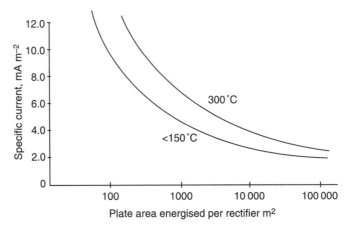

Figure 4.12   Effect of sectionalisation on specific current production

for plants operating at temperatures around 300 °C, assuming that the discharge electrodes are optimally spaced and the plants are operating just at the point of flashover. In practice, although the specific corona current decreases with the total plate area energised, for current power plant installations having individual field sizes of at least 5000 m² the rate of reduction is not too significant and hence the impact on overall performance is limited.

The corollary to the above is, if an existing installation is not meeting the required performance level, then it may be possible to improve the efficiency by splitting the field into smaller bus sections, each energised with its own rectifier equipment.

It will be noted from Figure 4.12 that, for very small bus sections, such as those met on pilot plants, as a result of the higher specific current and hence potential power input, the performance in terms of migration velocity is much higher than one would obtain from a large installation. Depending on the equipment supplier's philosophy on pilot testing, some use a scale factor when considering pilot data, whereas others reduce the specific corona input to equate to the full scale consumption. Thus, providing the pilot data is backed by full scale performance results, either approach can give satisfactory design information, with the proviso that the pilot testing must be carried out under the same inlet conditions, e.g. gas composition, temperature, dust concentration and particle sizing, as for the full scale plant.

4.7 High tension insulators

Although early installations had what would now be considered poor high tension insulators for supporting and insulating the discharge frames from the casing, modern ceramic materials, such as high alumina porcelains, alumina, silicon nitride, silica, etc., are currently used. Two approaches are employed to support the discharge frame, either directly vertically loaded, as in the case of the 'flower pot' design, or as outboard post insulators, where the discharge frame dead load is carried from a cross structure. In either case it is important that the insulator is maintained in a purged hot dry condition, such that deposition by any flue gas component and resultant electrical tracking is avoided. These approaches have more or less superseded unpurged lead through insulators, where the insulator acted only as a gas seal, the discharge system being carried from outboard dead post insulators. Generally, the requirements for any support insulator is that the material must have sufficient strength to support the weight of the discharge electrode system and be electrically strong enough to withstand operational voltages of up to 110 kV in the case of 400 mm spaced collectors, without breakdown or subjecting the rectifier equipment to electrical loading because of surface leakage.

Table 4.2 identifies some typical ceramic insulating materials and their mechanical and electrical properties [7].

Although the electrical resistivity of these materials is high at ambient temperatures, as the temperature increases the resistivity decreases as indicated in Figure 4.13 [8]. This figure, in addition to the above materials, also includes data on quartz and silicon nitride for comparison purposes. These curves indicate that, for operation at elevated temperature, the choice of material is limited if electrical leakage is to be minimised. If this leakage is associated with contamination by the flue gas components, this could result in severe arcing and subsequent failure of the insulator. In the case of porcelain insulators although the base material has good electrical properties, some of the glazes used for waterproofing, etc., are salt based and the sodium ions can act as charge carriers, particularly at higher operational temperatures, which may preclude their usage.

Dependent on the application and operating temperatures, most of the above materials can be used in an appropriate form and size for discharge electrode system support insulators. For discharge electrode rapping drive insulators, high

Table 4.2 Properties of some ceramic insulating materials

Property	Porcelain	90% Al_2O_3	99.5% Al_2O_3
Water absorption	None	None	None
Compressive strength	1930 MPa	2482 MPa	2620 MPa
Flexural strength	269 MPa	303 MPa	358 MPa
Coefficient of linear expansion	$5.3 \times 10^{-6}\,°C^{-1}$	$6.1 \times 10^{-6}\,°C^{-1}$	$7.1 \times 10^{-6}\,°C^{-1}$
Dielectric strength (1.27 mm thick sample)	17.3 kV mm^{-1} ac	17.7 kV mm^{-1} ac	16.9 kV mm^{-1} ac
Dielectric constant	8.2	8.8	9.8
Volume resistivity at 25 °C	$>10^{14}$ Ωcm	$>10^{14}$ Ωcm	$>10^{14}$ Ωcm

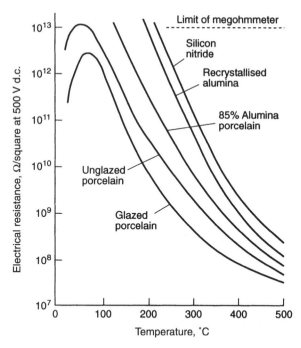

Figure 4.13   The effect of temperature upon the electrical resistance of insulating materials

density alumina porcelain material, having improved mechanical properties to withstand the rapping transmission loads, is normally employed.

4.8 Electrical clearances

The electrical clearances that must be carefully considered, if the precipitator is to operate efficiently and as designed, fall into two main categories: the first are those associated with the discharge electrodes and collectors; and the second, which are sometimes overlooked, are the peripheral clearances outside the field itself. Any electrical clearance that is too small will limit the operating voltage of the unit and so affect the efficiency of precipitation.

In general, the electrodes are suspended so that they hang exactly mid-way between adjacent collectors. In practice, although the alignment is critical, with erection and fabrication tolerances, one must accept a misalignment of up to ±10 mm on field heights of 10 m and above, but a tighter tolerance for smaller plants, which usually have higher design migration velocities.

The position of the discharge element within the collector and the distance between elements is important to maximise efficiency; generally, a minimum electrode element electrode element distance of half the collector/collector spacing is used to give satisfactory corona formation and current distribution over the collector surface (see Chapter 3). For collectors, the clearance between the stiffening channel edges and any emitting point should be greater than the collector half-spacing by some 25 per cent to provide satisfactory emission and voltage conditions. This distance applies in both the horizontal and vertical planes, since the electrodes or collectors may not hang exactly plumb.

Another area of concern with the discharge electrodes is the position where any support or carrier members pass through the collectors. In this case, not only should the member be low emitting, e.g. have a large radius of curvature, but must also pass centrally between adjacent collectors.

For clearances outside the actual field area, the positioning of any live discharge electrode support members must be well clear of any earthed structural components. Generally for safety, the following 'rule of thumb' minimum clearance distances should be adopted for any design. As positively energised components have a lower breakdown strength than negatively energised elements, any sharp edges on the earthed collector, casing or support members can result in premature breakdown. For example, for components having sharp edges opposing one another, a separation of three times the discharge electrode/collector distance should be used. Even with reasonably curved surfaces, such as those on rolled angles or joists, the minimum separation should be 1.5 times the discharge electrode/collector distance, with intermediate separations being adopted for differing profiles and edge conditions.

A special area of concern is where the discharge electrode support tubes pass through the casing and insulator. To minimise the quantity of heated purge air used with the cone or flower pot insulator, the open area at the base is significantly reduced by a restrictor, which would take the form of a profiled electrical stress cone or bus ring. This, with a large diameter suspension tube, e.g. 50 mm or so, effectively minimises the risk of breakdown in this area, in spite of the actual clearance approaching, or being sometimes less than, the discharge electrode/collector separation. The location of the suspension tube must be truly central within the bus ring to be fully effective.

Where bus ducts, rather than HT cable, are used to feed power to the electrodes, their diameter and profile must be such as to eliminate the risk of breakdown within the duct itself. If insulators are employed to centralise, or carry the bus connection, these should have adequate tracking length to prevent breakdown resulting from atmospheric dust deposition, etc.

On large precipitators having discharge electrode frames spanning two or more hoppers, the actual distance between the frame and any cross-hopper apex member should take into account any potential dust deposition on the member to minimise the risk of breakdown.

Similar comments would apply to internal tumbling hammer discharge electrode rapping systems, where the 'live' shaft and hammers could give rise to suspect areas, which would be difficult and expensive to overcome.

In general the following clearances are recommended for any design of precipitator.

Condition	Minimum clearance
Rounded corner on DE system to flat earth	1.5 × ½ duct width
Sharp edge on DE system to flat earth	2.0 × ½ duct width
Sharp corner on DE system to earthed sharp corner	2.5 × ½ duct width

Other components, which may be built into the system, should adopt the above 'rule of thumb' minimum clearance distances in order that the maximum applied voltage is across the discharge electrode/collector system and not elsewhere in the plant.

4.9 Deposit removal from the collector and discharge systems

In modern dry installations, dependent on the supplier, the rapping is achieved either by a rotating tumble hammer or by dropping a heavy rod onto an anvil connected to the component being rapped. The tumble hammer system is normally operated through a motor-driven shaft and the drop rod by either a motor-driven camshaft or a magnetic lifting mechanism. Older systems, employing vibrators or mechanical lifting and dropping of the collectors, have been superseded by the foregoing, but can sometimes be found.

In order that the particles reach the hopper, the size of the agglomerate for power plant fly ash, when dislodged, must be some 1000 μm in effective diameter to overcome the horizontal component of gas flow [9]. This means that a compromise has to be established on any operating plant, such that (a) the rapping provides sufficient energy to dislodge the deposited layer from the surface without causing the agglomerated layer to shatter and become re-entrained, and (b) also limits mechanical damage to the components being rapped. Because the rate of deposition varies along the precipitator, the rate or frequency of rapping is reduced from the inlet to the outlet fields. This is necessary to enable the deposited layer to achieve sufficient thickness to produce agglomerates of sufficient size to reach the hoppers when sheared from the surface by the rapping.

Although the frequency of collector rapping is relatively slow, it nevertheless is mechanically searching; at a rate of 12 raps per hour the components undergo one million impact cycles in approximately 10 years. Therefore, to eliminate premature failure most suppliers carry out full scale laboratory investigations. These are not only to ensure that the system is capable of long term reliability, by equivalent 10 year life testing, but also that the energy distribution is satisfactory and reasonably uniform across the collector or discharge system.

4.9.1 Impact of collector deposits on electrical operation

A representation of the dust layer on the collector plate is shown in Figure 4.14. In practice this layer usually consists of a permanent bonded layer resulting from repeated plant start ups and shut downs when the temperature passes

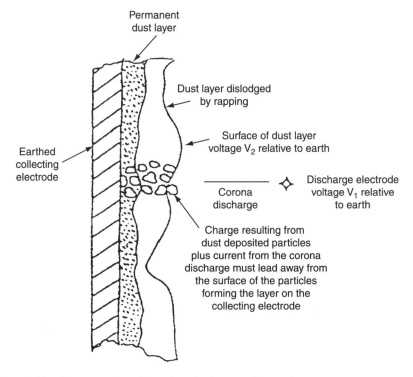

Figure 4.14 Representation of dust deposited on a collector plate

through the dew point, covered by a layer of loose relatively recently deposited dust. Only part of this layer will shear off when rapped, i.e. dust is only dislodged from dust and not the **metal surface**. The presence of the bonded layer further increases the required thickness of dust on the collector plate necessary for the release of optimum agglomerate sizes in order to reach the hoppers.

The presence of this dust layer can seriously impact on the electrical operating conditions, dependent on the thickness and the resistivity of the dust itself. There are generally three modes of electrical operation of precipitators, for example:

- normal favourable dust – typical full scale corona current flow around 10×10^{-5} A m^{-2};
- resistive dust – corona current typically restricted to 3×10^{-5} A m^{-2};
- high resistivity dust – reverse ionisation – positive corona current flows from plates to discharge electrodes in addition to the negative flow. The two currents are additive and normal metering will not discriminate between them. Reverse ionisation is a complex phenomenon and does not really lend itself to the approach used for the other two conditions where there is only negative corona discharge.

Assuming the dust acts as a series resistance, the voltage drop across the dust layer on the collector plate as illustrated in Figure 4.15. For a dust layer of 1–3 mm thickness the voltage drop is insignificant for resistivities below 10^{11} Ωcm, and even for a thickness of 5–10 mm the drop would be only about 1 kV even with the higher corona current possible with favourable dusts.

In the case of highly resistive dust, in excess of 10^{12} Ωcm, the voltage drop across 3 mm of dust becomes significant, even with a restricted corona current flow. Since there is likely to be a significant residual bonded layer, the total thickness for the release of agglomerates at least 1 mm in size can readily approach or exceed 3 mm.

This might appear to be an over-simplification of the effect of the dust layer but appears to have some credence. For example, in Figure 4.16 are shown voltage–current curves for a favourable dust C, a high resistive dust B and a high resistivity dust A resulting in reverse ionisation. Note that the measured electrode voltages, i.e. the breakdown voltages, for dusts B and C are approximately the same, but the control level for the high resistive dust gives less than half of the corona current. The kilovolt meter measures the voltage from the discharge electrode to earth, but the precipitation field voltage is actually between the discharge electrode and the surface of the dust layer. The two are only equal when the voltage drop across the dust layer is small as with favourable resistivity dusts – below 10^{12} Ωcm.

The difference in voltage between B and C for the control setting of B is approximately 7 kV. Assuming this is the voltage drop across the dust layer, then the precipitator effective field voltage is roughly 80 per cent of that for the favourable dust and the corona current is about 40 per cent. Subtracting 7 kV would put the operating level of B close to that of C, giving voltages more representative of the actual value. The lower effective field voltage, and consequent lower current, could readily explain the lower performance of B

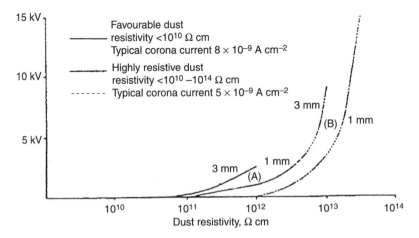

Figure 4.15   Calculated voltage drop across dust layers for high and normal resistivity dusts

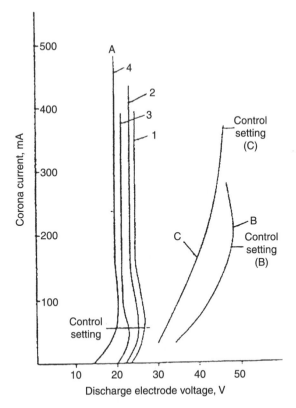

Figure 4.16   Comparison of electrical characteristics for varying resistivity dusts

compared with C. Practical evidence supports the view that for a favorable dust resistivity, i.e. less than 10^{12} Ωcm, the thickness of deposits on the collector system in general has relatively little effect on operating conditions [10].

Although the greatest rate of dust deposition occurs on the collectors, even small accumulations on the discharge electrodes can have a deleterious effect on the electrode emission characteristics. Therefore, it is equally, if not more, important to ensure that the discharge electrodes are effectively rapped, since the radius of curvature would be increased thereby reducing the available corona current (Chapter 2). As the discharge system contains very much less material than the collectors, transference of energy through the discharge system is more difficult. Measurements have shown that some 50 per cent of the total applied energy is lost through every 90° change of direction; nevertheless, both internal tumble hammer and external drop rod rapping systems are employed. The internal tumble hammer raps either an individual frame or a pair of frames, whereas the external drop rod form of rapping can only impart energy to a specific area of the support frame. By limiting the size of individual support frames, very effective rapping can be achieved, albeit at the cost of additional insulators and support frames.

In the case of a dry precipitator, the usual method of collector deposit removal is by means of some form of mechanical impact rapping; however, the very act of rapping can result not only in the risk of particle re-entrainment, but, as the dust is disturbed, in electrical breakdown, both of which will impact adversely on the performance. To minimise the possibility of this, some designs reduce the electrical operating voltage by a few kilovolts just prior to the hammer falling, while others rely on a fast acting automatic voltage control (AVC) system to minimise any resulting upset condition. In other plants, power-off rapping is employed; i.e. all electrostatic clamping forces are removed to assist in dust dislodgment.

Other installations attempt to control re-entrainment and electrical upset by completely isolating the flow during rapping by means of inlet and outlet isolating dampers, so-called off-flow rapping; all fields are then rapped simultaneously, usually with the power off. The isolated flow is then left for some period of time to allow the dust to reach the hoppers prior to reconnecting the flow back into service. An alternative to complete flow isolation is to blank-off only the duct being rapped with an air curtain or damper system [11]. The economics of these approaches need careful evaluation, since the cost of the additional dampers and duct seals may outweigh the gain from a potentially reduced plate area precipitator.

With wet type applications, there is little risk of the deposit, on either the collector or discharge system, resulting in electrical breakdown. It is important, however, in the case of the wash down approach, where the TR equipment is isolated during the wash down period, that sufficient time elapses prior to re-energising the TRs to avoid electrical breakdown resulting from any remaining liquor dripping from the internals. For continuously irrigated units it is imperative that the design incorporates drainage facilities to collect and lead the washing liquor safely away from the collectors and discharge electrodes, otherwise electrical breakdown and shorting, which will significantly affect performance, can arise.

4.9.2 *Measurement of rapping intensity on dry precipitator collectors*

It is generally accepted in the precipitation industry that, if rapping intensities of at least 100g (zero to peak) can be achieved on the collectors and a similar value on the discharge electrodes, then the resultant cleanliness will prove satisfactory in terms of maintaining performance. While 'g', the acceleration due to gravity, is used as a measure of rapping effectiveness, this is very much an oversimplification, since the acceleration value determined will be dependent on the size and thickness of the element being rapped as shown by the following formula [12]

$$\text{acceleration} = 4\pi^2 f^2 d,$$

where f is the fundamental frequency of vibration (Hz) and d is the displacement (m), and

$$f = \frac{K}{2} \times \left(\frac{Dg}{wA^4}\right)^{\frac{1}{2}},$$

where K is a constant, $D = Et^3(1 - V^2)$, t is the plate thickness, V is Poisson's ratio, E is the modulus of elasticity, w is the weight per unit area and A is the short edge dimension.

Because the measurement of g to determine the effectiveness of rapping is not the complete story, various manufacturers often quote lower values than the above $100g$ for their particular system as being adequate to enable their plant to perform satisfactorily. In this instance, one must consider the reputation and track record of the supplier, prior to accepting the statement covering lower g values, which may be correct for the vast majority of particulates met in practice by that particular supplier.

Ideally, since the collector is relatively thin, the measuring device should have zero mass to eliminate local disturbances and changes in the mode of plate vibration. In practice, a lightweight $0.2g$ accelerometer is currently the nearest approach to the ideal and to avoid 'ringing' and therefore falsely high readings, some suppliers advocate that the practical approach is to use an equipment measuring cut-off frequency of 10 kHz. This frequency is well below the collector's natural frequency of vibration, but to date there is no practice or procedure that has been adopted world-wide to assess these measurements.

4.10 Hopper dust removal

Examination of the data logs from many operational plants indicates that at least 70 per cent of the reported dry precipitator problems are associated with hopper dedusting difficulties of some sort, so one cannot overemphasise the importance of developing a satisfactory hopper and dedusting system for any precipitator installation.

There are numerous devices on the market for hopper level indication/detection, employing one or more of the following principles: capacitance change, change in a tuning fork frequency, change in pressure/suction, change in rotational speed of a driven disc, ultrasonic and radioactive sensors. As any device will determine only a local dust situation, overfull hoppers can still arise, in spite of an apparent 'empty' hopper indication. Should the hopper overfill with dust, not only will the dust short out the electrical supply but could also result in severe mechanical damage if the level should reach high enough into the field area. During emptying of a very overfull hopper, if 'rat holing' occurs and only one duct empties, the hydrostatic pressure exerted by the relatively fluid fly ash on adjacent partially filled ducts can result in distortion of the lower part of the collector with a deleterious effect on precipitator performance. Sometimes the discharge system itself can also be damaged or twisted, with similar impact on the electrical operation and collection efficiency.

To minimise the above or at least to give further early warning, in addition to the normal hopper level devices, flexible metal chains are sometimes hung from the four corners of the discharge frame. The dust on reaching the chains usually causes the electrics to trip on reduced kilovolt output, thereby preventing further dust being precipitated. One should remember, however, that although dust will not be electrically precipitated, within the inlet field in particular it can still deposit under gravity and hence create a problem. Therefore, action to remedy an overfull hopper situation must have a high priority if potential damage to the internals, with the attendant significant fall in performance, is to be avoided.

4.11 Re-entrainment from hoppers

As many precipitators operate under suction, it is imperative that air inleakage into the hopper region is prevented to eliminate possible re-entrainment of already collected particulates. Even an 'unmeasurable' inleakage flow can re-entrain sufficient dust in passing through the hopper area to significantly affect the overall emission from the plant. This is particularly important with the outlet hoppers because there is no possibility of further precipitation occurring, for example, assuming a precipitator designed normally to operate with an emission of 50 mg Nm^{-3}, will, if there is hopper inleakage equating to only 0.25 per cent of the total flow but carrying 10 g Nm^{-3} of re-entrained dust, produce a final emission approaching 75 mg Nm^{-3}, i.e. a 50 per cent increase.

For volumetric systems, e.g. conveyors of various types, rotary valves are often used sited below the hopper take-off point to meter the dust flow into the conveyor. As the rotary valve acts as a pump on the return empty half-cycle, its position in the down leg and operation must be such as to provide a positive head of dust to eliminate inleakage. For either dense or lean phase pneumatic conveying systems, as both rely on slide-type gate valves for isolation and sealing, the above 'pumping' problem does not arise; however, it is most important that these gate or slide valves effectively seal, particularly when positive pressure systems are used.

4.12 References

1 GRIECO, G. J.: 'Electrostatic Precipitator Electrode Geometry and Collector Plate Spacing – Selection Considerations'. Proceedings Power-Gen 94 Conference, Orlando, Fl., USA, 1994, Paper TP 94-58
2 LODGE, O. J.: 'Improvements in Means for the Production of Continuous High Potential Electrical Discharges Applicable for Deposition of Dust, Fume, Smoke, Fog and Mist for the Production of Rain and other Purposes'. British Patent No. 24,305, October 9th 1903
3 COTTRELL, F. G.: 'Art of Separating Suspended Particles from Gaseous Bodies'. US Patent No. 895,729, August 11th 1908
4 COTTINGHAM, C. R., and PARKER, K.R.: 'An Electrostatic Precipita-

tor Designed Specifically to Collect Orimulsion Fly Ash'. Proceedings 10th Particulate Control Symposium and 5th ICESP Conference on ESPs, Washington, DC, USA, 1993; EPRI, Palo Alto, Ca., USA, TR-103048 V2, Paper P7-1

5 HOULGREAVE J. A. et al.: 'A Finite Element Method for Modelling 3D Field and Current Distribution in Electrostatic Precipitators with Electrodes of any Shape. Proceedings'. 6th ICESP Conference. Budapest. Hungary. June 1996. pp 154–159.

6 WHITE, H. J.: 'Industrial Electrostatic Precipitation'. (Addison Wesley Pub. Co. Inc., Pergamon Press Ltd, Reading, Ma./Palo Alto, Ca./London, 1963

7 COORS Porcelain Company: Technical Bulletin No 953

8 LODGE COTTRELL Ltd.: Unpublished Research Report

9 LOWE, H. J., and LUCAS, D. H.: 'The Physics of Electrostatic Precipitation'. *Brit. J. of Applied Physics*, 1953, Supplement No 2

10 BAYLIS, A. P., and RUSSELL-JONES, A.: 'Collecting Electrode Rapping Designed for High Efficiency Electric Utility Boiler Electrostatic Precipitators'. Proceedings 4th EPA/EPRI Symposium on the *Transfer and Utilization of Particulate Control Technology*, Houston, Tx., USA, 1982; EPRI Publications, Palo Alto, Ca., USA

11 FALAKI, H. R.: 'Experimental Study of Flow Diversion during Rapping of Collector Plates inside an ESP'. Proceedings 10th Particulate Control Symposium and 5th ICESP Conference on ESPs, Washington, DC, USA, 1993, pp. 37.1–15; EPRI TR-103048 V2, Palo Alto, Ca., USA

12 DARBY, K.: 'An Examination of the Full Electrostatic Precipitation Process for Cleaning of Gases'. Proceedings 10th Particulate Control Symposium and 5th ICESP Conference on ESPs, Washington, DC, USA, 1993, pp. 26.1–17; EPRI TR-103048 V2, Palo Alto, Ca., USA

Chapter 5

Development of electrical energisation equipment

5.1 Early d.c. energisation techniques

The first commercial application of electrostatic precipitation was that installed at the lead smelter of Parker, Walker and Co. at Baguilt in North Wales. This plant resulted from Lodge's investigations and findings from his work at Liverpool University in 1884 [1]. This repeated in more detail earlier work by Holfeld (1824) [2] and Guitard (1850) [3] when it was found that particles of smoke could be precipitated from a smoke filled vessel by applying a charge to the particles by means of a Voss or Wimshurst electrostatic generator.

Walker recognised the potential of Lodge's work to reduce emissions from his lead smelter complex, and applied for and received a patent [4] describing the physical arrangement of the plant. While the unit comprised an arrangement of insulated discharge elements mounted within the ductwork, the electrical energisation was from a modified large Wimshurst machine consisting of two 1.52 m (5 foot) diameter discs driven from a 1 HP steam engine. The electrostatic voltage developed by the machine was stored in large capacity Leyden Jars before being applied to the discharge system. The plant unfortunately failed to produce the anticipated results because it was quickly recognised that although the voltages were high enough to result in ionisation, the capacity of the Leyden Jars was insufficient to maintain the necessary corona/ion flow to charge the particles for effective precipitation.

Following this setback it was recognised that the effect was not truly electrostatic and that the equipment had to supply corona current in order to charge the particles. At the end of the nineteenth century, most electrical supplies were direct current (d.c.), rather than the later alternating current (a.c.) supplies. At this stage, the principle of transforming a low d.c. voltage into a much higher voltage was, however, well established in the form of the Ruhmkorff Induction Coil, illustrated in Figure 5.1.

The principle of operation of the Ruhmkorff Coil is basically a step up transformer, the primary of which is fed from an interrupted d.c. supply, derived from

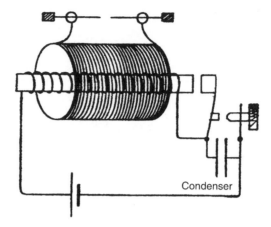

Figure 5.1   Principle of the Ruhmkorff Induction Coil

a mechanical spring loaded spark gap. At a certain magnetising current the spark gap opens and the magnetic field collapses, the spring loading then remakes the spark gap connection and the process is repeated. The collapse and remaking of the magnetic field in the primary results in inducing a voltage in the secondary windings, which, depending on the turns ratio, permittivity of the core, etc., produces an interrupted high voltage wave form of the type indicated in Figure 5.2. The self-inductance of the primary winding on 'make' reduces the flux rate of change, while on 'break' the winding resistance is practically infinite which makes the break much faster, producing the higher rate of flux change and consequently a high voltage output. It is normal practice to mitigate arcing of the contact points by adding a condenser connected across them such that the back e.m.f., owing to the self-inductance of the primary, is reduced.

Although investigations were made by Lodge and other workers using the Ruhmkorrf Coil, the practical problems associated with sparking and burning

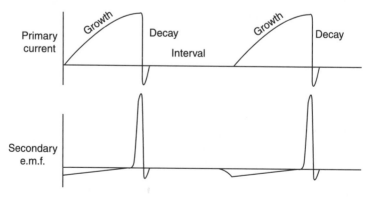

Figure 5.2   Effect of self-inductance of primary upon output waveform

of the contacts and the short duration voltage, and hence power pulses, obtained from the induction coil limited its application in the precipitation field.

5.2 Development of a.c. mains frequency rectification equipment

5.2.1 *Valve rectifiers*

It was not until the beginning of the twentieth century that significant developments occurred in the electrical equipment field, when electric motors, dynamos, improved insulating materials, etc., became commercially available leading the way towards improved and more reliable d.c. power supplies.

During this period, Lodge (in conjunction with Robinson) investigated the use of valves for rectification purposes. The first equipment was based on the use of mercury arc rectifiers, in a Cooper–Hewitt bridge arrangement, connected to the secondary of a large Ruhmkorff Induction Coil supplied from a gas driven d.c. dynamo source [5]. This work was then followed in 1905 by a vacuum diode developed in conjunction with Muirhead and Robinson for X-ray applications using an a.c. supply source [6], which Lodge subsequently used for precipitation duties. Figure 5.3 indicates details of the valve, while Figure 5.4 shows a transformer house using the valves.

As an alternative to the mechanical switched rotary rectifier, which will be reviewed in Section 5.2.2, the continued development of diode valves led to their being used almost universally at this time for electrostatic precipitation duties. Although the earlier valves had cold cathodes, later valves using heated cathodes and other improvements in material deposition, e.g. using thorium and other coatings having good emission characteristics, supplanted the cold cathode for precipitation duties. When these hot cathode diodes were used for high voltage rectification, it was necessary for the secondaries of the heater transformers to be insulated against the high voltages. Developments in the transformer and insulation material fields made this approach to high voltage rectification possible. Circuit diagrams showing valves used for both full and half wave rectification are illustrated in Figure 5.5 (a and b), indicating the use of a centre tapped transformer to reduce the number and hence cost of the rectifiers required.

A further high voltage arrangement using diode valves, was the voltage doubler, as shown in Figure 5.6. Although the output is d.c., the use of the doubler circuit, employing capacitors, meant that the second stage resulted in only around 50 per cent of the current being available at the output terminal. Although this limited its use in precipitation duties to energising relatively small plate areas, the 'doubler' approach is widely used to obtain very high d.c. voltages by increasing the number of stages in series, each stage roughly doubling the potential output voltage.

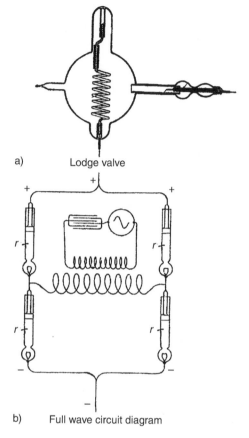

a)   Lodge valve

b)   Full wave circuit diagram

Figure 5.3   Lodge high tension rectifier valve [7]

5.2.2 *Mechanical switch rectifier*

At the time of Lodge's involvement with valve rectifiers, parallel work was being carried out in the United States, under the direction of Dr F. G. Cottrell, in developing a mechanical rotary rectifier system for precipitation duties [7]. Initially the precipitators were to counter sulphuric acid mist discharges from the plants around San Francisco and later the metallurgical smelter fumes at Anaconda.

This form of rectification was for many years to be almost universally used throughout the precipitation industry as one of the major sources of energisation. Although the device was extremely robust and maintenance was usually limited to periodically changing the contacts, because the switching took place in air, not only did the rectifier have to be installed in a specially constructed switch room for safety reasons, but the arcing at the contact points resulted in the formation of ozone and nitrogen oxides. The formation of these gases, which are regarded as hazardous and hence falling under the auspices of the Health

Figure 5.4   Photograph of an early transformer house using Lodge equipment [7]

and Safety Executive, led to mechanical switch rectifiers being finally abandoned in favour of static rectifier devices.

An early photograph of a rotary rectifier is shown in Figure 5.7 [8], which shows the HT transformer output connected to insulator mounted shoes positioned adjacent to the opposite arms of a motor driven rectifier, which rotates at half the mains supply frequency.

The principle of operation of the mechanical switch rectifier is indicated in Figure 5.8. At the positive peak of the alternating voltage waveform, a break down connection is made to the earth terminal of the set, while the negative peak is connected to the output terminal. When the frequency moves forward by 180°, the motor and arms only move through 90°, so the arms are in the correct position to connect the positive peak to earth and the negative peak to the outlet terminal, thereby providing a negative output. (Note that if the motor phase is shifted through 180° the output will be positive.)

In later developments, the two rotating arms shown above were replaced by a disc, fabricated from an electrical insulating material having metallic edge connections over the two opposite 90° sectors of the disc, as shown in Figure 5.9. This photograph also indicates the control panel, HT transformer, primary

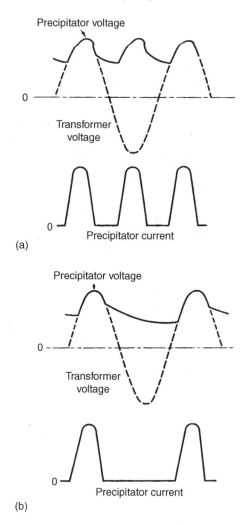

Figure 5.5 (a and b)   Circuit diagrams showing full and half wave rectification

auto-transformer and primary resistor; these components will be discussed later in the chapter.

   To ensure that the rotating connection arms were in the correct phase position, the early machines had an adjustable position stator, later versions had adjustable disc location facilities and finally the motors had a permanent magnet to ensure correct phasing. Although negative energisation is typically used for industrial precipitation duties, the rotary rectifier can produce either positive or negative voltages dependent on the phasing of the motor in relation to the frequency of the a.c. supply.

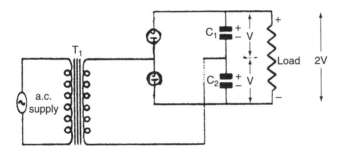

Figure 5.6   Voltage doubler circuit arrangement

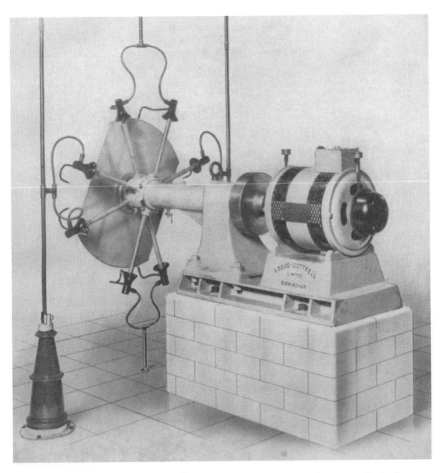

Figure 5.7   Mechanical switch rectifier contacting arrangement (courtesy Oxford University Press)

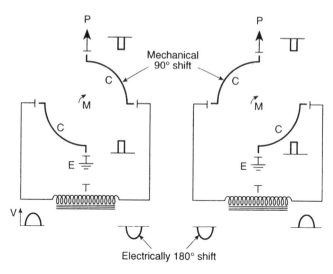

Figure 5.8   Principle of operation of mechanical rectifier

Figure 5.9   Later development of rotary rectifier arrangement (courtesy Lodge Sturtevant Ltd)

5.2.3 *Metal oxide rectifiers*

Almost in parallel with the usage of the diode valve, metal oxide rectifiers were being developed in the 1920s and were later used for precipitation duties. These devices, having the advantage of being more robust and not requiring a high voltage heater transformer, were to prove more cost effective and better suited for long term reliability. Commercially in the early work, two types of material were found to exhibit rectification properties: copper oxide and selenium.

The initial rectifiers were based on copper oxide, where a 1 mm thick disc of pure copper was coated on one side with a 0.1 mm layer of cuprous oxide deposited by thermal treatment. The cuprous oxide, so formed, is very hard and offers a low resistance to the passage of current flowing from the copper oxide to the copper but a very high resistance in the opposite direction from the copper to the oxide. Figure 5.10 indicates a single layer rectifier, with a counter electrode, formed from a lead disc supported on a graphite coating on the copper oxide surface to enable the electrical connections to be made, the whole assembly being held under pressure. It is postulated that the all important asymmetry to current flow is the result of the 'barrier layer', which exists between the cuprous oxide and copper.

A later development was based on a selenium layer deposited onto a nickel coated steel disc. The outer layer of selenium was then coated with a low melting point tin alloy to form the counter electrode, which when held under pressure makes a mechanically sound practical assembly. Figure 5.11 illustrates the construction of such a rectifier.

In practice, the forward resistance of the deposited layer is not zero and the backward resistance is not infinite (Figure 5.11); therefore, during operation

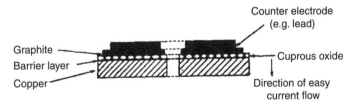

Copper oxide rectifier element (passes conventional current $Cu_2O \rightarrow Cu$).

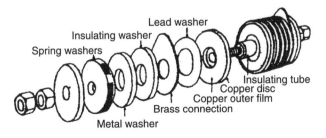

Figure 5.10   Mechanical arrangement of a copper oxide rectifier

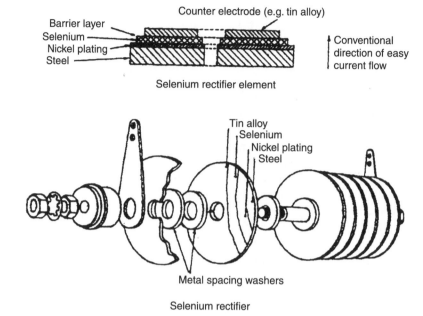

Figure 5.11   Mechanical arrangement of a selenium rectifier

considerable heat is developed with either rectifier arrangement. If adequate cooling is not provided, the resultant increase in device temperature will cause significant changes in resistance and hence its electrical characteristics and rectification performance. Excessive temperatures increases will result in permanent damage to the rectifier itself. Typical working temperatures are limited to 30 °C with a maximum temperature rise of 40 °C, as indicated in Figure 5.12.

5.2.4 Silicon rectifiers

Virtually all current types of electrical energisation equipment utilise silicon diode rectifiers in their design. Not only are the diodes smaller, but they usually have a much higher working voltage and smaller voltage drop than the foregoing metal oxide types, making the installation more compact. Although germanium exhibits similar properties and was initially considered for precipitation duties, the superior reverse voltage and current characteristics of silicon, being around 1/100 of those of germanium, together with its having limiting operating temperatures about double that of germanium, meant that silicon became the preferred device for most rectification duties.

The development of the silicon rectifier dates back to the 1960s, when a slice of 'n-type' mono-crystalline silicon was diffusion coated with boron, through a small window to form a region of 'p-type' material. The abrupt change where the boron/silicon interface occurs is termed a 'p-n' junction, leads being subsequently connected to both types of material to form the diode itself. Although,

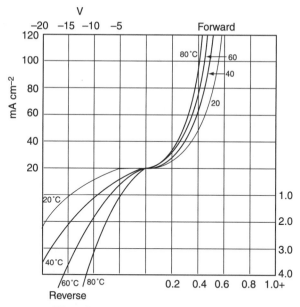

Effect of temperature on copper oxide rectifier

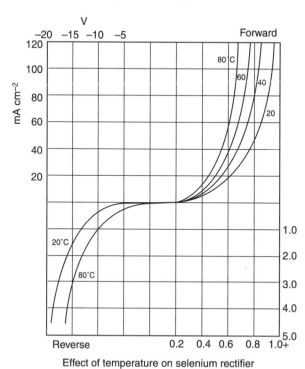

Effect of temperature on selenium rectifier

Figure 5.12   Rectifier characteristics with varying temperatures

initially both the p- and n-type materials are electrically neutral, when energised, the applied voltage results in 'holes' being able to diffuse through the p-type material, while electrons cross through the n-type material into the p-type. The p-type, in losing 'holes', gains electrons and hence acquires a negative charge, while simultaneously the n-type gains a positive charge.

Under equilibrium conditions, the acquired negative charge on the p-type material prevents further electrons crossing the junction and conversely the positive charge prevents further 'holes' crossing. The resultant space charge creates an internal barrier across the junction. Because of this barrier, the region close to the junction is free of majority carriers and is termed the 'depletion layer'. The induced voltage across the junction is inversely linked to the intrinsic carrier concentration, N, which for silicon is $1.4e^{10}$ cm^{-3}, giving a junction voltage of around 600 mV.

In terms of operation, under forward biasing conditions, the holes move towards the cathode and the electrons towards the anode, while at the junction the holes and electrons combine to form a conductive path providing the diode with a low resistance. When under reverse bias, the holes tend to migrate towards the anode, while the electrons move towards the cathode, resulting in a 'depletion layer' having insulating properties. In essence, the diode has a conduction path under forward bias conditions and a non-conductive path under reverse biasing (see Figure 5.13).

Being relatively small compared with metal oxide types of rectifier, two forms of diode assembly can be used in the design of transformer rectifier equipment, both comprising a series string of diodes typically mounted on printed circuit (PC) boards. These PC units are usually connected as a bridge arrangement and positioned adjacent to the transformer within the oil filled containment tank. With one arrangement, the diodes are typically '1000 V' devices connected in series, each having a parallel capacitor and resistor network to ensure voltage sharing between each diode. The resistor enables the reverse bias voltage to be shared, while the capacitor effectively distributes high frequency voltages arising

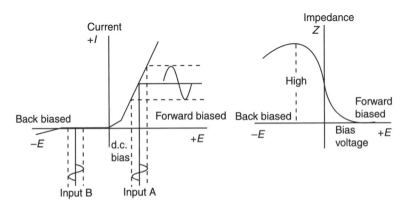

Figure 5.13   Illustration of a typical conduction diagram for a silicon diode

from sparking, etc., to be dissipated. The alternative arrangement employs 'controlled avalanche' diodes, which have typical reverse breakdown voltages in the order of 1500 V or higher. These devices, although not requiring resistor/capacitor sharing devices, must be reasonably matched in terms of avalanche voltage to ensure voltage sharing. If the voltage sharing is not uniformly distributed among the devices, those with the lowest leakage will accept a disproportionate share and the diodes will fail in the classical 'domino' fashion. For either arrangement the rating of the bridge/diode arm must have a peak inverse voltage (PIV) of at least twice that of the transformer maximum peak voltage to mitigate potential failure resulting from 'ringing' within the system. Ringing results from the system's capacitive and inductive components, usually following flashover, producing a resonant oscillatory circuit, having a few kHz frequency, that can develop very high voltages which can appear across the bridge arm circuitry.

Although silicon devices have some toleration against excessive temperature operation, if the safe value is exceeded, through poor heat transfer within the containment tank or as a result of excessive arcing within the precipitator, the junction itself can fail and the whole voltage appears across the junction causing arcing. If the arcing is allowed to continue it will ultimately carbonise the oil producing conductive carbon particles, which can lead to complete failure of the equipment, and/or result in an explosive gas mixture being found in the roof area of the transformer tank.

Rectification can be either full wave, which produces continuous voltage and currents coinciding with the mains frequency, or half wave, using a centre tapped transformer, where the voltage and current waveforms coincide with alternate pulses of the mains frequency. Generally with present precipitator applications having large plate area fields in order to achieve the required emission levels at an economic cost, it is more usual to find the full wave approach being used.

A typical full wave circuit and the corresponding wave forms are indicated in Figure 5.14.

5.3 HT transformers

Transformers traditionally used for precipitation duties are of a conventional design as used for normal mains supply operation, composed basically of primary and secondary windings wound on a laminated core, with appropriate electrical insulation between the core, primary and secondary windings.

Two winding arrangements can be found on transformers used for precipitation duties, either as separate coils wound on opposite sides of the core (core wound) or concentric with the secondary being wound on top of the primary (shell wound). In the case of the core wound approach, the secondary coils are often 'pancake coils' having an output of around 4 kV stacked together as opposed to a continuously wound coil. For either arrangement it is important

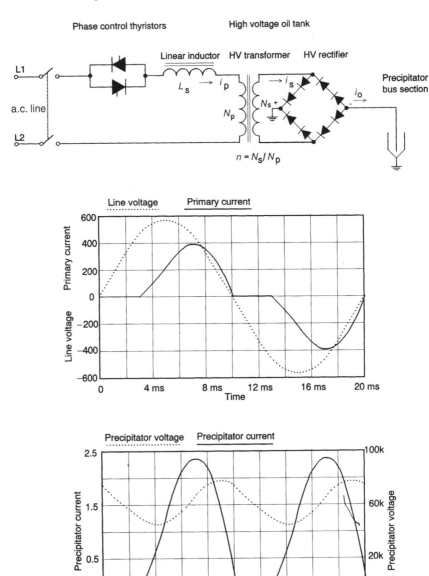

Figure 5.14   Full wave circuit diagrams and waveforms

that the final output windings on the secondary are 'double spaced' to minimise stress failure as a result of 'ringing' produced from flashover within the precipitator. The windings typically comprise cotton or shellac insulated copper wire with additional insulation placed between adjacent layers and between the core and first winding. The coils themselves must be capable of accepting repeated sparking and arcing within the precipitator fields, since these can result in current surges well above the nominal system rating. These surges can also exert considerable electrical and mechanical stresses on the windings and system insulation and unless accounted for can result in premature failure of the equipment.

The principle of operation is that an alternating supply fed to the primary windings sets up an alternating current and hence a changing magnetic flux. This alternating flux field is linked to the secondary winding through the core and hence a voltage is induced across the windings. This secondary induced voltage is then after rectification used to energise the electrostatic precipitator.

For precipitation duties the following features need to be considered in the design of the transformer.

(a) The necessary turns ratio of the windings: for 400 mm collector spacings the operating voltage of the transformer should be rated at around 80 kV d.c., which means for a mains on a 415 V a.c. supply, the turns ratio will be around 190:1. (For smaller collector spacings, the operating voltages are lower and the turns ratios are reduced.)

(b) The peak secondary voltage determines the insulation requirements and internal clearances within the equipment. For a 400 mm spaced precipitator the transformer equipment secondary peak rating is ~113 kV.

(c) The maximum temperature rise permissible as determined by ambient conditions.

(d) The current rating necessary to produce sufficient corona current to ensure the particles are fully charged during their passage through the field.

(e) The maximum voltage available to the primary windings allowing for impedance losses.

(f) The kVA rating, which will determine conductor sizes, core size and cooling requirements.

5.3.1 Transformer losses

Although modern construction and materials mitigate the various losses associated with the transformer, those that should be considered are as follows.

5.3.1.1 Iron losses

As the inductance of the primary is not infinite, current will flow through the primary when the secondary is 'off' load. During operation it is the magnetising current that provides the necessary flux density and although it is constant for all loads, it is 90° lagging with the voltage and consequently must be considered as a loss. This alternating flux, in addition to producing the transformed voltages in the secondary windings, also results in voltages being developed as eddy currents in the core material. Owing to hysterisis losses occurring during the magnetising of the core material, these appear in the form of heat; consequently both eddy current and hysterisis losses combine as a heating effect and form a resistive component to the supply.

Hysterisis losses may be minimised by the correct choice of core material, that is one having a small area B/H loop and laminating the material to reduce eddy current losses. The main flux within the transformer varies only slightly between no load and full load operation and may be considered constant over the normal operating range. The hysterisis loss can be determined by measuring the input power to the transformer when the secondary is open circuited.

5.3.1.2 Copper losses

These losses are simply caused by the d.c. resistance of the windings and can be represented as I^2R, that is, the losses rapidly increase with load current, hence the correct selection of wire diameter, etc., can significantly impact on the losses which appear as heat. The copper loss at full load operation can be determined by short circuiting the secondary and applying a small input voltage to the primary of sufficient value to produce full load secondary current. Under these conditions, as the iron loss is small compared to the copper loss, a wattmeter across the primary can be assumed to measure only the copper losses.

5.3.1.3 Flux leakage

In a perfect transformer all the flux produced by the primary cuts the secondary windings; in practice, however, some of the flux will also cut the primary windings, which results in the primary having a degree of self-inductance, because of the leakage.

5.3.1.4 Self-capacity of the windings

Although at mains frequency this effect is not too important, it can be significant at higher frequencies, such as those used in switched mode power supplies, which will be reviewed in Chapter 8.

In practice the losses associated with the transformer can be readily measured with the aid of an ammeter and wattmeter connected across the primary. To measure the copper and core losses, the secondary must be open circuited and the primary voltage raised to its maximum. An ammeter connected in the primary determines the copper loss, which for a good transformer is less than 5 per cent of the full load current, and the I^2R loss at no load is ~1/400 of that at full load, while the wattmeter reading gives the total core loss of the transformer, which is usually constant with load and represents a loss of ~2 per cent. For full load tests the secondary is short circuited through an ammeter and the primary voltage is slowly increased to achieve the full secondary current output.

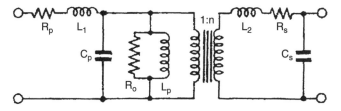

Figure 5.15   Equivalent circuit diagram of a transformer

Assuming the primary I^2R is the same as that for no load operation and the core loss is small, then the wattmeter gives the I^2R of the transformer.

The equivalent circuit of a transformer is complex and is represented in Figure 5.15.

A correctly designed and operated transformer will give a life well in excess of 20 years and is probably one of the most reliable pieces of equipment used in the precipitation industry. Failures can, however, occur. The first type of failure is through a degenerative mode, usually attributable to defects in the materials used in the construction; this is typically associated with the insulation materials, which fail through over stressing, excess temperature or mechanical movement of the components forming the transformer. The second form of failure can occur through the transformer being subjected to excessive stressing of the windings either by over voltages or currents, which can produce a significant movement within the windings and usually result in the foregoing degenerative mode of failure.

5.4 Cooling of transformer and rectifier equipment

Because of the various losses associated with the transformer and rectifiers resulting in the liberation of heat, it is essential that adequate cooling of the unit be provided to dissipate this thermal effect, otherwise the life of the unit will be compromised. On most installations, the transformer and rectifiers are contained in an oil filled vessel, either as a breathable arrangement with a reservoir tank, or in the form of a completely sealed unit, having an inert gas atmosphere above the oil level in order to accommodate changes in volume. These arrangements, depending on the output capacity and losses, may have cooling fins or a separate air cooled radiator system, the quantity of oil and the cooling facility being assessed to limit the temperature rise to around 40 °C.

In addition to providing a ready method of cooling, the dielectric, insulating and thermal transference properties of the oil means that these forms of construction are much more compact than an air insulated unit. The oil used for transformer rectifier equipment can be mineral, silicone based (poly-dimethyl siloxane), or high molecular weight hydrocarbon based fluids (HTH). The latter, although having reduced cost and lower flammability characteristics than the mineral oils, have a higher viscosity which may present the designer with thermal

Table 5.1 Comparison of the characteristics of insulating oils

	Mineral oil	Askarel	Silicone	Midel 7131
Viscosity cSt at 25 °C	16	15	50	100
Density, g ml^{-1}	0.88	1.5	0.96	0.98
Flash point, °C	150	195	300	257
Vol. resistivity, Ωcm	$1 \times e^{14}$	$1 \times e^{12}$	$1 \times e^{15}$	$1 \times e^{13}$
DDF at 25 °C, 50 Hz	0.0005	0.0005	0.00002	0.001
Permittivity at 25 °C	2.2	4.5	2.7	3.2

transfer difficulties at lower ambient temperatures. A comparison of the different oil characteristics is illustrated in Table 5.1.

The oils/dielectrics used for precipitation and other duties have to meet stringent requirements in order to provide a satisfactory equipment life. These requirements are met using standard test procedures; in the UK, tests are conducted under British Standard methods, while in the US, ASTM procedures are adopted. No doubt other standards are used elsewhere; Europe for instance will have a VIM or equivalent Standard. These requirements and tests can be summarised as follows.

(a) Neutralisation number

During operation the oil may come into contact with various substances, which may result in its contamination. In general, the test is only applied to oil extracted from equipment that has been in use for a period of time and is basically used to determine the acidic content of the oil. The data are then used for comparison against fresh oil and other test data, to indicate if the oil needs to be changed/cleaned before serious damage to the equipment occurs.

(b) Dielectric breakdown voltage

For any insulating material, the breakdown voltage is used to determine its ability to withstand electrical stress without failure. The test measures the breakdown voltage between two electrodes under specific test conditions and its value indicates if contamination from moisture, dirt or conductive matter is present in the sample, which will reduce its capacity as an insulator.

(c) Interfacial tension

This test determines the interfacial pressure between the oil and any water present in the oil. It actually measures the molecular attractive forces between unlike molecules at their interface and is a means of detecting soluble contaminants and products of deterioration of the oil or its additives.

(d) Power factor

This is a measure of the power dissipated in the oil in watts compared to the effective voltage × current value, when tested with a sinusoidal supply under specific test conditions. A high value indicates the presence of contaminants, e.g. products of oxidation, dirt or charged colloids in the oil, which will impact on its suitability.

(e) Colour

This test identifies the value of transmitted light through a sample compared to a series of colour standards. A high colour number is an indication of deterioration of the oil and/or the presence of contaminants.

(f) Water in oil (Karl Fischer)

Contamination of the oil can be in the form of suspended free droplets or dissolved water in the oil. Free droplets are readily identified by visual examination/cloudiness, whereas dissolved water is more difficult to assess and the Karl Fischer apparatus is used to determine the level of dissolved water in terms of parts per million (p.p.m.). This test is important since the electrical dielectric strength test may not identify the presence of water, which may lead to ultimate failure.

Table 5.2 identifies the values, determined from the tests, as being satisfactory for long term operation of the equipment.

5.5 Primary input control systems

5.5.1 Manual methods

The earliest system for controlling the input voltage was simple manual tap changing on the transformer primary feeding the high voltage unit. The control achieved by this system was rather coarse and time consuming, because the precipitator had to be electrically deenergised during the connection change;

Table 5.2 Acceptable test results for long term operation

Test method	BSS 142 1972	ASTM (method)	
Neutralisation number	Max 0.4 mg KOH g^{-1}	Max 0.4 mg KOH g^{-1}	(D664)
Dielectric breakdown voltage	30 kV (2.5 mm gap)	22 kV min.	(D877)
Interfacial tension	n.a.	18 dynes DM^{-1} min.	(D971)
Power factor	n.a.	1.0% Doble limit	(D924)
Colour	n.a.	4 max.	(D1500)
Water content	50 p.p.m. max.	55 p.p.m. max.	(D1533)
Viscosity	40 cSt at 20 °C	n.a.	
Sludge	0.10% max.	n.a.	
DDF	0.005 at 50 Hz	n.a.	
Flash point, °C	140 min.	n.a.	

consequently the unit was rarely optimised and the operators left the unit on a lowish output level to minimise their involvement.

The manual change over was subsequently replaced by a switch system, but again, since it was manually controlled, the TRs tended to be left in a 'safe' trouble free system by the operators hence optimum performance was rarely attained in practice.

5.5.2 Motorised methods

The development of the motor controlled auto-transformer or 'Variac' resulted in the operators being able to control the input supply voltage remotely by simply energising the motor control. The auto-transformer or Variac comprises a primary winding having a motor driven movable contact running along the winding surface, so that a continuously variable input voltage can be tapped off to feed the fixed secondary winding.

At the initial stage of application of this approach, precipitator instrumentation was rather basic, usually consisting of a voltmeter across the HT transformer, a primary ammeter and a secondary milliammeter connected across a shunt resistor in the earth leg of the rectifier circuit. As regards control, the primary voltage was manually increased until flashover was detected, then the voltage 'backed off' by some set amount to minimise circuit tripping. Unlike other electrical equipment that rarely experiences a flashover/breakdown situation, that used for precipitation duty is always seeking the maximum voltage applicable to the electrode system, that is, just at the point of electrical breakdown, hence the equipment must be designed to cater for this condition.

A precipitator is electrically represented as a 'leaky' capacitor, that is, the electrode system forms the capacitor and the corona current the resistive component. The self-capacitance of a reasonably sized precipitation field can be some 40 pF and hence at an operating voltage of say 60 kV, the total charge stored within the system ($q = CV^2$), is some 144 J, which can be dissipated during a flashover. In the early equipment, in order to limit the primary current rise, a resistance of around 0.5 Ω was typically included in the primary, but could be adjusted up to some 2 Ω dependent on the operating conditions experienced by the precipitator.

A typical circuit using a motorised Variac with line current limiting resistors is illustrated in Figure 5.16.

To eliminate spark erosion and wear on the moving contact of the Variacs, moving coil regulators were used as an alternative to control the voltage input. In one arrangement, the primary transformer feeding the HT transformer comprises two windings, the secondary of which can either move horizontally along the axis of the primary or rotate within the primary such that the magnetising flux can be changed depending on the position of the secondary. Otherwise the operation of the system is similar to the circuitry illustrated in Figure 5.16.

On later equipment designs, in order to minimise current surges during flashover and arcing and simultaneously reducing power losses, the total circuit

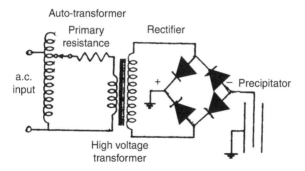

Figure 5.16   Typical circuit diagram using a Variac and primary resistors

impedance is arranged to be around 40 per cent, with 5 to 10 per cent being in the primary windings of the transformer, the remainder being in added as an inductor, typically having an inductance between 5 and 20 mH. The 40 per cent total impedance limits the primary current rise to around 250 per cent of the design line current of the transformer. Although higher values of total impedance would reduce the current rise, the time lost in re-establishing the optimum voltage on the precipitation field can significantly increase, with the resultant fall in precipitator performance during the recovery time.

5.5.3 Saturable reactors

An intermediate development was the current limiting saturable reactor, which was used prior to the significant changes resulting from the development of silicon based devices in the 1960s and 1970s leading to the current usage of thyristors or silicon controlled rectifiers (SCRs) as primary control devices. The principal advantage of the saturable reactor or magnetic amplifier over the earlier forms of control is the ability to control high levels of power by the application of small control signals.

The saturable reactor consists of an inductor, with a split core, having a secondary d.c. winding, the current through which determines the total flux developed and hence the inductor's overall impedance. Its operation, therefore, depends on the variation of the inductance of the iron cored coils due to the level/change of magnetic flux within the core. A circuit diagram using this form of controller is illustrated in Figure 5.17.

During operation, the saturable reactor relies upon its inductive reactance in order to control the TR input power, but in doing so various difficulties can be experienced. Firstly, the saturable core reactance response is very slow in respect to changes in the d.c. energising current settings; and secondly, the creation and collapse of the magnetic field in the saturable core takes a relatively long time when compared with the standard 10 ms half line cycle. Several sparks may occur before impedance levels can be increased sufficiently to quell the undesirable element of any sparking. Once established, an arc can persist for a longish

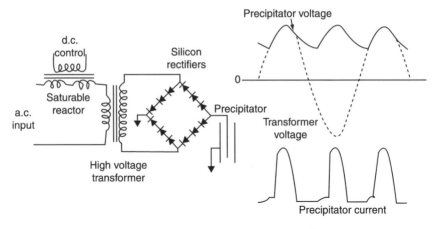

Figure 5.17   Typical circuit employing saturable reactor control

period of time, which can not only result in damage to the precipitator internal elements, but during this period the performance will be compromised.

The second problem with the saturable core reactor is that it is virtually impossible to shut the TR input current down entirely. The impedance of the saturable reactor is minimal for zero d.c. magnetising current, increasing with rising current flow until magnet saturation is achieved. This minimal impedance value can result in a serious 'switch on' problem, because without d.c. current flow, the maximum line voltage is directly applied to the HT transformer producing maximum output voltage which will lead to severe flashover within the precipitator. This situation, however, can be avoided by delaying the actual closure of the contactor feeding the HT transformer until the d.c. circuit is fully energised and supplying current to the control winding of the saturable reactor.

Although the saturable reactor can produce a satisfactory control system for most applications, if a 'soft arc' develops then the controller will continue to feed current into the arc at around 150 per cent of the maximum line current even at maximum d.c. energising current. If this condition is encountered in practice the only way of returning the unit to normal operation was to isolate/ disconnect the primary feed to the HT transformer for a short period of time in order to dissipate the arc.

5.5.4 Silicon controlled rectifiers (SCRs)

The silicon control rectifier or thyristor is shown schematically in Figure 5.18. Its operation makes use of the breakdown of a reverse biased p-n junction contained within two outer layers referred to as the anode and cathode. The arrangement and characteristic curve is also illustrated in Figure 5.18.

The arrangement indicates that both the anode and cathode junctions are forward biased, but the control junction is reverse biased. Under this condition the current through the device is made up of electrons in the p-n control layer

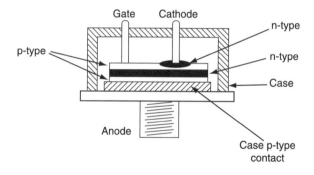

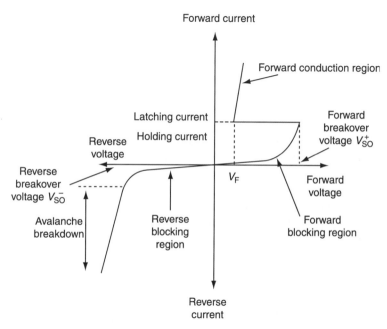

Figure 5.18   *Thyristor arrangement (top) and characteristic curve (bottom)*

diffusing towards the anode and the 'hole' carriers in the n-type blocking layer migrating towards the cathode. Initially the current is fairly low and represents the saturation current of the reverse biased control junction; as the voltage across the device is increased, the current rises until the breakover point is reached. At this point the reverse field is strong enough at the control junction that avalanche breakdown occurs, allowing further current to pass such that the overall resistance of the device drops sharply. The potential difference across the device correspondingly falls and 'kicks back' to the normal low conduction characteristic. This breakdown condition is reversible and the original state may be obtained by reducing the applied voltage until the current falls below the holding current.

The junction breakdown can be induced earlier by applying a positive potential to the control layer with respect to the cathode so that the reverse field is increased. The control layer is then called a 'gate' and an electron current flows into the control layer and hence to the anode. Breakdown can now occur at a much lower voltage and the current through the device is independent of the 'gate' potential and can only be reduced by reducing the anode-cathode voltage.

A typical thyristor 'turn-on' relationship between the gate current and the gate cathode voltage is illustrated in Figure 5.19. This, in addition to indicating maximum and minimun gate voltages and currents, also illustrates the effect of the 'turn-on' characteristics under differing temperature conditions.

For a transformer rectifier primary voltage controller, two thyristors are connected in an anti-parallel configuration, as indicated in Figure 5.14. One thyristor controls the positive half cycles and the other the negative half cycles. Both thyristors are balanced in that the gate 'switch-on' voltages are similar and are protected against dv/dt damage by a snubbing resistor/capacitor network; they are also typically fused against current overload. In operation, two conditions must simultaneously apply, the first is that the devices must be forward biased and the gate voltage must be adjusted to provide a 'trigger' signal. By varying the time delay between when these conditions are satisfied, the amount of time the thyristor is conducting varies; this variation is referred to as the thyristor conduction angle, which can range from zero, 'thyristor off', to full 180°, i.e. full firing. Typical waveforms exiting the thyristor assembly are also illustrated in Figure 5.14, from which, it can be seen that the r.m.s power in the primary circuit is controllable by adopting different firing angles.

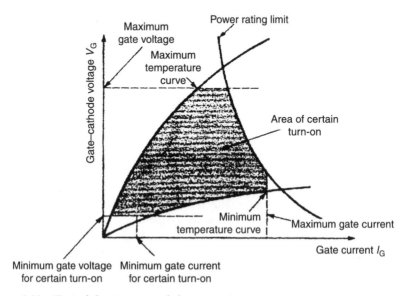

Figure 5.19   Typical thyristor control characteristics

The majority of operational precipitators now use silicon controlled rectifiers or thyristors for control of the primary power and their usage will be covered in greater detail in Chapter 6.

5.6 Automatic control systems

5.6.1 Early current control

With the rapid increase in industrialisation following World War II and the tightening of particulate emission regulations, it became important to be able automatically to optimise precipitator performance by operating the precipitation field close to the breakdown voltage at all times. The first attempts utilised the rapid increase in primary current resulting from flashover and a simple 'ramp–detect–back off' approach was developed. Although an improvement over manual adjustment of the voltage, to minimise wear on the moving contacts of the auto-transformer then in common usage for primary voltage control, the ramp rate had to be fairly slow, with a 30 s response time being typical. The overall system therefore lacked inertia to control the precipitator in order to achieve optimum performance.

Figure 5.20 indicates how the mean operating voltage on the precipitator changes with flashover rate. This shows that for a specific application, the optimum precipitator voltage is achieved with a flashover rate of around 120 per minute. Once the optimum sparking rate has been reached any further increase in primary voltage results in a decrease in mean operating voltage and hence efficiency at the expense of increasing power usage because of 'active' time lost through arcing. Although for the particular application, Figure 5.20 indicates an optimum sparking rate around 120 per minute, for different operating conditions the optimum rate may differ being basically site/application dependent.

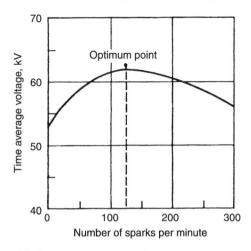

Figure 5.20   Effect of flashover rate on mean voltage on precipitator

With the change from auto-transformer to saturable reactor for primary control, the 'ramp–detect–back off' approach enabled the system to have an improved response time of some 50 ms. This meant that the precipitator voltage could be maintained closer to the breakdown condition with changes in operating conditions within the precipitator field and hence achieve improved overall performance levels.

5.6.2 Voltage control method

Since the precipitator performance is proportional to the operating voltage squared (see Equation (2.14)), it became important to develop a means of monitoring the operating voltage on the discharge electrode system. This was achieved by series connecting 1 MΩ resistors to produce a string having a total resistance of around 100 MΩ This string was connected directly to the HT line and the lower end connected to earth through a 30,000 Ω resistor in parallel with a micro-ammeter; this approach readily enabled the average electrode voltage to be evaluated/measured. Because of the large time constant of the circuit, only average values can be measured using a resistor chain; for instantaneous and peak voltage measurements the voltage divider must be of a capacitance type.

[*Note.* To protect personnel and instrumentation against potential high voltages, the voltage micro-ammeter and secondary current milli-ammeter must be connected across substantial shunt resistors connected to earth.]

Figure 5.21 indicates the relationship between input voltage, precipitator voltage and emission for a power generating plant precipitation field, which supports that the minimum emission, i.e. optimum collection efficiency, is coincident with the maximum precipitator operating voltage. This led to a

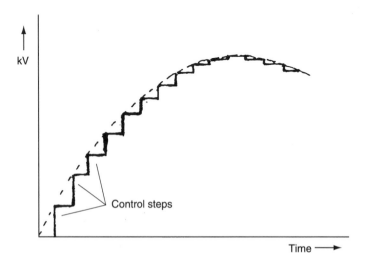

Figure 5.21   Principle of hill climbing form of controller

system of 'hill climbing' being developed as an alternative to the 'ramp/fall back' approach; here, instead of rapidly ramping the primary voltage up to flashover, the primary voltage was steadily raised in steps until flashover was achieved, at which point, the control system lowers the primary voltage by one or more small, 0.5 kV equivalent controllable steps, depending on the basic operating conditions to eliminate the flashover. After a certain time period, the raise system is again initiated which slowly lifts the voltage up to flashover and the whole control system cycles about the maximum flashover voltage in a controlled fashion.

There have been numerous discussions within the precipitation industry as to which system results in the best precipitator performance; however, under ideal operating conditions both can produce similar efficiencies and there is little to choose between them. The 'hill climbing' approach, however, tends automatically to select the optimum operating voltages, whereas, the 'ramp/fall back' approach may need manual adjustment to optimise the sparking rate.

Although either approach can result in better overall performance levels than manual forms of control, it is important that the AVC operation produces a certain degree and intensity of flashover. This is necessary to maintain the discharge electrode emitting points clear of deposit, otherwise the true breakdown voltage tends slowly to 'fall back', which results in a lower than optimum efficiency. Site investigations have indicated that unless the control point corresponds to a similar voltage value as that obtained from the V/I 'clean plant' characteristic, then the AVC can operate, apparently satisfactorily, at a voltage some 3–4 kV lower than optimum, which results in a significant fall in corona current. (A recent test indicated that by increasing the flashover rate and intensity resulted in an additional 200 mA being supplied to an inlet field, with corresponding increases in the downstream fields as a result of reduced space charge effects.)

To minimise reflected transients arising during flashover, which can impact on the circuit components, particularly the rectifiers, the output line connection is typically fitted with heavy duty line control resistors or high voltage inductors of some type. Following a flashover there is always a significant 'inrush' current to recharge the capacitive element of the precipitator, since during the breakdown the time constant of the precipitator is almost zero being a capacitor discharge. For installations employing high voltage cable rather than direct bus connection, these line control resistors or secondary inductors are essential. Depending on the length and total capacitance of the cable and other circuit components, the induced voltages resulting from 'ringing' can be at least twice the nominal operating voltage of the precipitator and have been known to result in cable failure.

5.6.3 Computer control methods

Most operational plants presently employ computer based automatic voltage control systems. These, dependent on the programme and using electronic switching, have response times in the order of milliseconds which overcomes the in-built mechanical limitations of the earlier approaches.

With the rapid development of computers during the late 1970s and early 1980s, leading to the production of A/D converters, analogue readings from primary and secondary ammeters and voltmeters can be produced as digital signals and directly used in specially adapted computer programmes. These programmes tend to be specific to any one manufacturer, and since each manufacturer will have a preferred method of control philosophy, it is only possible to consider general principles of operation. These approaches will be reviewed in Chapter 6, but are mentioned here to complete the section on component and energisation system developments.

5.7 Summary of developments

Over the past century the electrical energisation of electrostatic precipitators has seen many changes, from the initial demonstrations using static voltage generators through to full silicon based equipment, most systems in current usage adopting transformation and rectification of the incoming a.c. supply voltages and frequencies for energisation purposes. In the past decade, however, with the availability of faster high power switching devices and developments in high frequency transformers, etc., there is now a trend to adopt switch mode power supplies operating at frequencies of some 25 kHz. The development and operation of these new devices will be detailed in Chapter 8.

Generally rectifier developments have gone hand in hand with advancements made in electrical and electronic engineering, starting with the diode valve, then the metal oxide rectifier through to the use of silicon diodes.

The initial control systems were largely mechanical based employing manual switching of the incoming voltage, then motorised versions, followed by magnetic amplifiers/saturable reactors and lastly by the use of silicon controlled rectifiers for transformer input control.

As regards optimising the precipitator performance by adjusting the incoming voltage to a level around breakdown, the initial approaches used simple manual control based on primary current, followed by improved mechanical arrangements as input control systems were motorised. These were followed by magnetic amplifier and analogue systems using analogue derived primary and secondary signals and finally by full digital methods employing computer programmes, using digital signalling derived from standard analogue metering devices.

Although the components used in precipitator energisation equipment have seen many changes over the past century, most equipment currently used for precipitation duties still basically comprises some form of incoming switch gear, a current limiting reactor, followed by a step up transformer and rectifiers which provide the outgoing d.c. voltage, the output line usually having some form of impedance to limit the impact of any flashover or arcing which may occur within the precipitator itself.

5.8 References

1 LODGE, O. J.: 'Dust Free Spaces'. Lecture to the Royal Dublin Society, 1884 (*see* Transactions of Society)
2 HOHLFELD, M.: 'Das Niederschlagen des Rauches durch Elekricitat. Kaustner Arch', *Gesammte Naturl.*, 1824, **2**, pp. 205–6
3 GUITARD, C.F.: 'Condensation by Electricity', *Mech. Mag.* (London), 1850, **53**, p. 346
4 WALKER, A.O.: 'Process for Separating and Collecting Particles of Metal or Metallic Compounds Applicable for Condensing Fumes from Smelting Furnaces and for other Purposes'. British Patent No. 11,120, August 9th 1884
5 LODGE, O. J.: 'Improvements in Means for the Production of Continuous High Potential Electrical Discharges Applicable for Deposition of Dust, Fume, Smoke, Fog and Mist for the Production of Rain and other Purposes'. British Patent No. 24,305, October 9th 1903
6 LODGE, O. J., *et al.*: 'Improvements to Vacuum Tubes for Electric Discharges'. British Patent No. 25,047, December 2nd 1905
7 COTTRELL, F. G.: 'Art of Separating Suspended Particles from Gaseous Bodies'. US Patent No. 895,729, August 11th 1908
8 LODGE, O. J.: 'Electrical Precipitation' *Physics in Industry*, 1925, **3**, Oxford Univ. Press, London

Chapter 6

Modern mains frequency energisation and control

Most modern precipitator installations are invariably energised from mains frequency rectified equipment utilising the latest electrical and electronic components as described in Chapter 5. Generally the transformer input conditions are controlled by anti-parallel connected thyristors/silicon controlled rectifiers, which in turn are controlled by some form of microprocessor based automatic voltage control (AVC) system. Although the supply is generally a fully rectified voltage, many modern equipment designs have the facility of intermittent energisation to cater for specific fly ash conditions, or for power saving while meeting a target emission. The basic operating characteristics of a mains frequency rectified power supply and the types of control will be examined in the following sections.

6.1 Basic operation of mains frequency equipment

The output voltage of the power supply, applied to a precipitator bus section, is controlled by varying the firing angle of the thyristors, i.e. by delaying or advancing the firing instant in relation to the zero crossing (crossover) of the line voltage. The basic circuitry of the electrical equipment is shown in Figure 6.1 for a conventional 50 Hz power supply.

When a thyristor fires, it changes from a state of high resistance to low resistance and the primary current starts increasing and becomes as shown in Figure 6.2(a). The magnitude and duration of the current is determined by the current limiting inductor (L_p), the leakage inductance of the HV transformer and the precipitator load. The output current (I_o), the precipitator current, has a waveform corresponding to the rectified secondary current of the transformer. The amplitude difference between the primary and the secondary current is related to the transformer turns ratio n, as $I_o = I/n$. The resulting precipitator current and precipitator voltage output waveforms are as depicted in Figure 6.2(b). These were obtained when the thyristors were fired at a time equal to 3 ms after the line

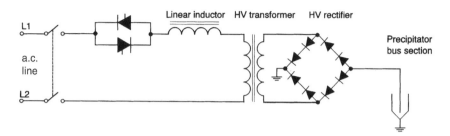

Figure 6.1   Typical mains frequency rectifier equipment

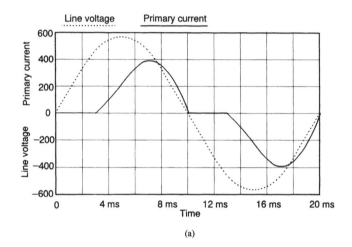

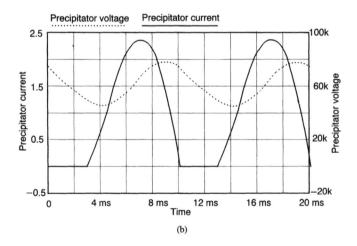

Figure 6.2   Voltage and current waveforms with a thyristor firing angle of 54°

voltage zero crossing, i.e. a firing angle $\theta = 180 \times 3/10 = 54°$. In practice these waveforms should be inverted, because of the negative corona, but for simplicity their absolute values will be used throughout this chapter and are typical for a medium size power supply energising a normal precipitator load.

The precipitator voltage shown in Figure 6.2 has a considerable ripple, because of the inherent capacitance of the precipitator section and the fact that no filtering is used. In general the precipitator voltage is characterised by its peak value (V_0peak); Mean value (V_0mean) and minimum value (V_0min). The peak voltage depends on the charge delivered in one current pulse (Q_p), i.e. the area beneath the current pulse. Because the precipitator capacitance is discharged in the time interval between the arrival of successive current pulses the value of the mean voltage decreases.

The voltage level can also be decreased by delaying the firing angle of the thyristors, which results in a lower precipitator voltage as indicated by the waveforms in Figure 6.3(a and b). These current and voltage waveforms illustrate the situation when the firing occurs 6 ms after the zero crossing point.

Table 6.1 presents as a comparison the operating values and their computer-aided design figures for thyristor firing times (angles) of 3 and 6 ms after the zero crossing point.

It is clearly seen that both the precipitator voltage and current are reduced when delayed firing is used. The extreme case corresponds to $t_0 = 10$ ms ($\theta = 180°$), giving a stationary current and voltage equal to 0.

The earliest practical firing of the thyristor is influenced by the value of the precipitator voltage at the firing instant that is equal to V_0min. This value is greater than the corona onset voltage and is influenced by the precipitator geometry and process conditions. With normal precipitator loads and a 50 Hz power supply, the earliest practical firing is in the order of 2–3 ms, corresponding to a firing angle of 36–54°.

Another important quantity, which must be considered, is the form factor (FF) of the precipitator current. This is defined as

$$FF = I_0 \text{ r.m.s.}/I_0 \text{ mean.} \tag{6.1}$$

The output current $I_0(t)$ is periodic and $|I_0(t)|$ is its absolute value.

Calculation of the form factor for a precipitator current when $t_0 = 3$ ms from Table 6.1 gives a value of 1.4, which is a typical design value for current mains frequency power supplies.

As the precipitator current is not a pure sinusoidal wave, it is not easy to theoretically calculate its true mean and r.m.s. values, however, they can be derived by computer simulation or by using approximations, as follows.

From the three Maxwell equations governing the electric field Poisson's equation can be expressed by the relationship

$$\nabla E = \rho/\varepsilon_0, \tag{6.2}$$

where E is the field strength (in V m^{-1}), ρ is the charge density (in C m^{-3}) and ε_0

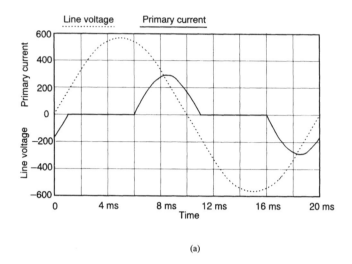

(a)

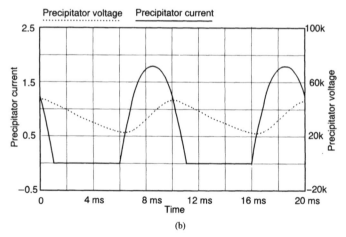

(b)

Figure 6.3   Voltage and current waveforms with a thyristor firing angle of 108°

Table 6.1   Current and voltages with two different firing angles

Firing angle	t_0 (ms)	3	6
Primary current	I_p(A)	223	145
Precipitator current	I_0 r.m.s. (mA)	1400	920
	I_0 peak (mA)	2350	1800
	I_0 mean (mA)	1030	576
Precipitator voltage	V_0 peak (kV)	78	47
	V_0 mean (kV)	61	35
	V_0 min (kv)	46	23

Courtesy FLS Miljö a/s.

is the permittivity of free space (8.85×10^{-12} F/ m^{-1}, which is valid for most gases under precipitator operating conditions).

The solution of Poisson's equation for a wireplate geometry is extremely complex, but it can be simplified if it is assumed that the current is small and the alteration of the potential by the ionic space charge can be represented by the same value theoretically derived for the simpler symmetrical wire-tube arrangement. The average current density, J_s, as a function of the potential at the discharge electrode can be expressed by

$$J_s = \frac{\pi \varepsilon_0 b}{cs^2 \times \ln(d/r_0)} \times V(V - V_c) \text{ (A m}^{-2}) \tag{6.3}$$

where b is the ion mobility (2.1×10^4 m^2 Vs^{-1} for negative corona in air), d is an equivalent cylindrical radius $d = 4s/\pi$ for $s/c \leq 0.6$ (s is half the collector separation and c is half the discharge element separation), r_0 is the wire diameter and V_c is the corona onset potential given by

$$V_c = r_0 E_c \times \ln(d/r_0) \text{ V.} \tag{6.4}$$

The corona onset field E_c, has been found empirically and for negative corona in air it is

$$E_c = \delta'(32.2 + \frac{0.864 \times 10^5}{(r_0 \delta')^{1/2}}) \text{ V m}^{-1}. \tag{6.5}$$

The relative gas density δ' is conventionally expressed in relation to 1 atm and 25 °C, i.e.

$$\delta' = 298/(298 + T) \times P_a, \tag{6.6}$$

where T is the temperature (in °C) and P_a the absolute pressure.

In practice, the mean and r.m.s. values of the ESP current can be derived using the following procedure. For the current waveform shown in Figure 6.2, the mean and r.m.s. values can be expressed by the following approximated equations:

$$I_0 \text{ mean} = 2(I_0 \text{ peak}/\pi) \times (\tau/T) \text{ and} \tag{6.7}$$

$$I_0 \text{ r.m.s.} = I_0 \text{ peak} \times (\tau/2T)^{1/2}, \tag{6.8}$$

where T is the half-cycle time, e.g. 10 ms for 50 Hz supply, and τ is the firing period.

6.2 High voltage equipment supply ratings

The high voltage power supply for an electrostatic precipitator is characterised by the transformer turns ratio (n) and the short circuit impedance present in the main circuit (X_{Lsc}). The impedance is predominantly inductive and is composed of the leakage reactance of the transformer and the linear inductor L, as

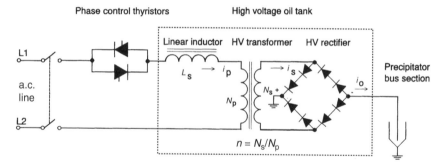

Figure 6.4   Typical impedance diagram of rectifier equipment

illustrated in Figure 6.4. This inductor L, is added in order to increase the short circuit impedance X_{Lsc} above the 5–10 per cent value, which is a typical figure for the leakage reactance of a standard HV transformer. The aim of the added impedance is to limit the current surges that might occur during sparking in the precipitator to

- increase the lifetime of the precipitator internals
- protect the electrical equipment, and
- obtain a more stable electrical operation.

The short circuit impedance value used by most of the equipment manufacturers is in the order of 30–40 per cent, which limits the current surges on short circuit to around 3/2.5 times the rated load current value.

By knowing the short circuit impedance/reactance, X_{Lsc}, and the transformer turns ratio n, all the important quantities of the power supply can be determined for different precipitator loads. The ratings can be expressed in different ways, but in this chapter the European practice will be used. The following quantities are normally indicated on the rating plate.

Precipitator mean current I_0 mean	mA
Precipitator peak voltage at no load (V_0 peak)	kVpk
Precipitator mean voltage V_0 mean	kV
Primary r.m.s. current (I_p r.m.s.)	A
Line voltage (V_L) and frequency (f)	V and Hz
Apparent input power (W)	kVA

6.2.1 Mean precipitator current

The precipitator current has a typical waveform as indicated in Figures 6.2 and 6.3. The mean secondary current value is given by

$$I_S \text{ mean} = 1/T \times \int_0^T I_0(t)\mathrm{d}t, \qquad (6.9)$$

where $I_0(t)$ is the precipitator current and T is the period of the line frequency.

The rated precipitator mean current is the maximum mean current the power supply is able to deliver to a load without exceeding the design primary current value. Some manufacturers carry out tests on the power supply with an RC-load, which simulates a precipitator load, while others use a pure resistive load, which gives a lower current form factor. This means that at rated mean load current, the r.m.s. value of the secondary and the primary currents have a different value compared with the corresponding figures using a precipitator load.

6.2.2 Primary r.m.s. current

The waveform of the primary current is shown in Figures 6.2(a and b). Its r.m.s. value is defined as

$$I_p^2 = 1/T \times \int_0^T I_p^2(t) \mathrm{d}t. \tag{6.10}$$

This can also be expressed as

$$I_p = n \times FF \times I_s \text{ nom}, \tag{6.11}$$

where n is the transformer turns ratio, FF is the form factor (typically 1.35–1.4) and $I_s$nom is the rated precipitator mean current.

6.2.3 Precipitator peak voltage under no-load conditions

At no-load, the output current of the power supply I_s is zero and the primary current is therefore equal to the magnetising current of the HV transformer. As this is negligible compared with the rated primary current, the peak voltage at no-load is equal to

$$V_0 \text{ peak} = \sqrt{2} \times nV_L \text{ nom}, \tag{6.12}$$

where V_L nom is the rated r.m.s. value of the line voltage.

6.2.4 Apparent input power

This quantity is especially important in the physical sizing of the electrical installation, cabling, switch gear, etc. The apparent input power S is defined as

$$S = I_p \times V_L. \tag{6.13}$$

That is, the product of the line voltage V_L and the primary current I_p.
    The active input power P is correspondingly given by

$$P = S \cos \varphi_1, \tag{6.14}$$

where φ_1 is the phase angle between the line voltage and the fundamental frequency component (50 Hz) of the primary current ($\cos \varphi_1$ is normally known as the power factor).

The active power cannot be expressed as a rated value, because it varies with the precipitator load for the same rated precipitator mean current. The active power is normally measured with a wattmeter or calculated by means of computer simulation. The power factor is normally better than 0.8 at full rated current; assuming the form factor is approximately 1.4.

6.2.5 Practical example

The waveforms depicted in Figure 6.2 were obtained with a power supply having the ratings given in Table 6.2.

Table 6.2 Rectifier equipment name plate ratings

Rated precipitator mean current	I_0	1200 mA
Rated precipitator peak voltage	V_0	110 kV
Line voltage	V_L	415 V r.m.s.
Short circuit reactance	X_{Lsc}	35%
Design form factor	FF	1.4

The precipitator is assumed to have collectors spaced at 400 mm centres and a calculated capacitance of 130 nF.

The transformer turns ratio can be derived as follows

$$n = \frac{1V_o}{\sqrt{2}V_L\text{nom}} = \frac{1}{\sqrt{2}} \times \frac{110\,000}{415} = 187. \tag{6.15}$$

The precipitator rated r.m.s. current I_0 is given by:

$$I_0 \text{ r.m.s.} = I_0 \text{ nom} \times FF = 1200 \times 1.4 = 1680 \text{ mA}. \tag{6.16}$$

The rated primary r.m.s. current is:

$$I_p \text{ r.m.s.} = n \times I_0 \text{ r.m.s.} = 18 \times 1.68 = 314 \text{ A}. \tag{6.17}$$

The rated apparent power is:

$$S = I_p \text{ r.m.s.} \times V_L \text{ r.m.s.} = 314 \times 415 = 130.4 \text{ kW}. \tag{6.18}$$

6.3 Influence of the linear inductor

6.3.1 The main function of the inductor

The linear inductor's main function is to limit the current surges arising during sparking within the precipitator, but, simultaneously, it also provides a significant number of advantages related to the precipitator's electrical operation.

To illustrate the disadvantages, Figure 6.5 indicates the precipitator current and voltage waveforms when the linear inductor is not used in the circuit, i.e. the current is only limited by the leakage reactance of the high voltage transformer.

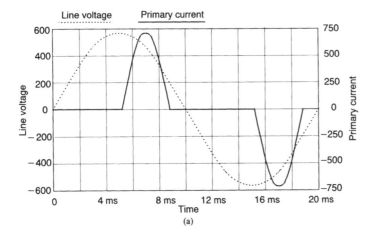

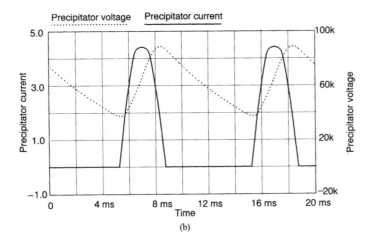

Figure 6.5   Precipitator waveforms without added impedance

The values obtained when the short circuit reactance of the transformer z is 9 per cent are shown in Table 6.3 in comparison to the same rated mean current and voltage as obtained with a linear inductor. (The figures in parentheses correspond to the values obtained with a normal short circuit reactance of 35 per cent.)

Comparing the results without added impedance to those obtained with a normal short circuit reactance (~40 per cent), Figure 6.2, for the same precipitator mean current and precipitator load, the following disadvantages become apparent.

(a) The peak value of the precipitator current is larger and its duration is shorter, resulting in a 36 per cent higher form factor.

Table 6.3 Current and voltages with insufficient short circuit impedance compared with values obtained with normal design impedance level (courtesy FLS Miljö a/s)

Firing angle	t_0 (ms)	5	(3)
Primary current	I_p r.m.s. (A)	302	(223)
Precipitator current	I_0 r.m.s. (mA)	1900	(1400)
	I_0 peak (mA)	4400	(2350)
	I_0 mean (mA)	1000	(1030)
Precipitator voltage	V_0 peak (kV)	90	(78)
	V_0 mean (kV)	61	(61)
	V_0 min (kV)	38	(46)

(b) The higher form factor results in a higher primary current and apparent input power for the same precipitator mean current.

(c) The time incidence of the precipitator voltage peak is closer to that of the line voltage peak value.

(d) Since sparking occurs around the peak of the precipitator voltage, the current surges have a higher amplitude and longer duration.

These characteristics are detrimental for stable operation of the precipitator, especially in relation to voltage recovery after spark.

In addition to the above the precipitator current, obtained with a delayed firing angle, produces a larger phase angle difference between the line voltage and the fundamental component of the primary current giving in a lower power factor, which is not desirable by the power supply companies.

However, the advantages of operation without adequate impedance can be theoretically claimed, as follows.

The precipitator voltage waveform is more pulsating because the peak value is higher, which could have a positive effect in the case of high resistivity particles. (Currently one system of resolving this problem is to use intermittent energisation or pulse charging techniques – see Chapter 7.)

In the case of precipitator loads requiring a higher voltage, the power supply is better suited to deliver its rated values because of the lower voltage drop across the short circuit reactance.

6.3.2 Physical implementation of the linear inductor

The linear inductor usually comprises an iron core inductor with a suitable air gap providing a linear impedance characteristic. This is normally placed inside the high voltage tank and its inductance cannot be changed. This is a good economic solution, provided a good match exists between the size of the power supply and the bus section.

Some manufacturers, however, locate the inductor inside the control cabinet, which can then be provided with changeover tappings. These allow one to change the inductance value to overcome any mismatch between the power supply and the operational bus section. This approach, however, has the inconvenience of occupying space in the control cabinet and causing acoustic noise during operation.

A variable inductor has now been developed and introduced where the inductance automatically changes inversely with the value of the primary current inside a certain range. This is achieved by means of an automatic control loop, which increases the inductance value when the primary current decreases in order to keep the form factor constant.

6.4 Automatic voltage control and instrumentation

Most investigators in this field have emphasised the importance of optimising electrical energisation in order to obtain the maximum collection efficiency. To accomplish this, two important aspects have to be taken into account.

(a) The size of the TR set and the energised bus section must be reasonably matched.
(b) The unit must have an efficient automatic voltage control (AVC) unit to optimise electrical operation under all operating conditions.

As already indicated, the key parameter governing the total corona power delivered by the TR set to a particular bus section is the firing angle of the thyristors. This angle is determined by the control unit for every half cycle of the supply frequency and must have the correct value according to the existing operating conditions within the precipitator.

The performance of any automatic control system is closely related to the type of instrumentation used to establish the precipitator conditions at any given time. The following section briefly reviews the signals typically used by the AVC units for control purposes.

The type of signal used for control purposes is typically dictated by the design approach used by different manufacturers and consequently are not always the same. The Europeans have a long tradition of using the actual precipitator current and voltage, the so-called 'secondary values', whereas the Americans have preferentially used the 'primary values', but in recent years the tendency has been to also incorporate the secondary values in their AVC units.

The use of r.m.s measuring devices along with high resolution analogue to digital converters enables modern microprocessor approaches to accurately sample the average operational voltage and current values of the precipitation fields. The actual parameters, which are used for control purposes to optimise performance in these later microprocessor based AVC units, are stored in a non-volatile memory using a key pad approach, and are not subject to change as could occur with earlier forms of controller.

Using the elements described above, coupled with a satisfactory control algorithm, it is possible to improve the controller response such that arcs within the precipitator are extinguished on a timely basis, thereby promoting improved precipitator performance. When handling difficult high resistivity dusts, etc., improved control techniques, e.g. intermittent energisation, can be used to mitigate any fall in performance as will be examined in Chapter 7.

Furthermore, the installation of opacity (or extinction) meters in chimneys is becoming more common, and in some countries is now compulsory, especially in connection with new and retrofitted plant. The signals delivered by these meters are used for continuous monitoring of the stack dust emission (CEM), but sometimes are also used by the AVC units for optimising power consumption. The purpose of the opacity meter in conjunction with the control units is to:

(a) optimise the operation of the precipitator to achieve maximum efficiency, and to
(b) achieve energy savings for a set emission/opacity under easy operating conditions.

The signals typically used by the AVC to control precipitator operation are depicted in Figure 6.6.

6.4.1 Secondary metering approach

In Europe, the TR equipment is normally fitted with secondary metering for measuring the precipitator voltage and current. The mean voltage is measured by a voltage divider, usually of the resistance type, and the current by means of a current shunt resistor connected in the ground or earth return connection of the rectifier. For the measurement of the peak and trough voltages some form of capacitive divider is used.

In modern AVC control units the following quantities are normally measured and displayed:

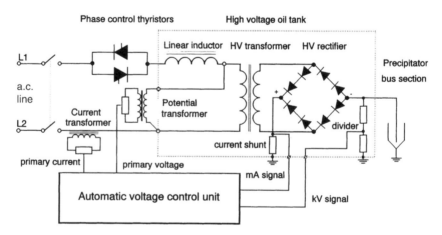

Figure 6.6   Signals used in the instrumentation and automatic control of a high voltage power supply

mean precipitator voltage,
peak precipitator voltage (peak of ripple voltage),
trough precipitator voltage (minimum of ripple voltage), and
precipitator mean current.

The mean values were initially used for control tasks but in the last decade, the importance of measuring the minimum or trough voltage value has become more significant [1]. This value is vitally important in evaluating the operation of a precipitator collecting high resistivity dust. Other important tasks using this feature are the automatic detection of back-corona and the control of the degree of intermittence, when operating under intermittent energisation [1, 2].

The peak voltage value, as previously indicated, is important in determining the optimum sparking level, i.e. maximum operating voltage in the precipitator and for optimising the electrical operation, e.g. during voltage recovery after a spark, and for computing the corona power.

In the past, the values of minimum and peak voltage were measured with an oscilloscope using a capacitance type voltage divider connected to the discharge frame unless one was actually built into the TR set. This measurement was, however, cumbersome for less experienced plant personnel, but today it is a standard feature in many modern AVC units using solid state integrated circuitry (ICs).

6.4.2 Primary metering system

In addition to the secondary values, primary input measurements are also used and displayed. These signals are used by some designs for automatic voltage control and/or monitoring tasks. It is now recognised, however, that the use of the secondary values is far superior for the automatic control and in the optimisation of the precipitator operation, e.g. spark detection, voltage recovery after spark and back-corona detection.

The primary values, however, can be used in various monitoring tasks, such as the determination of

- r.m.s. value of the primary current,
- r.m.s. value of the primary voltage,
- active power delivered to the TR set, and
- apparent power delivered to the TR set.

The primary voltage and current are usually measured by means of potential and current transformers connected into the primary. This approach enables the resultant outputs to be readily connected to the AVC unit, since the transformers provide electrical isolation and, dependant on the turns ratio, adequate signal levels.

The active power of the electrical equipment has two components:

(a)   the corona power delivered to the bus section, and

(b)   the losses in the TR set (transformer iron and copper losses, silicon rectifier diode conduction losses, etc.).

The active power delivered by the a.c. supply to the equipment has two main components:

(a)   the active power delivered to the TR set, and
(b)   the conduction losses within the thyristors.

Because the latter is negligible, compared with the corona power, the active power measured by means of the primary values is approximately equal to the active power delivered to the TR set. The equipment primary values are also important in monitoring tasks, such as

state of the phase control thyristors,
the value of the transformer current, i.e. over current,
the value of the transformer voltage,
magnetic saturation of the transformer, and
the value of the form factor, etc.

6.4.3   Opacity signal and full energy management

6.4.3.1   Energy saving system

In the past, certain regulatory bodies have raised objections to systems which target set point (as opposed to optimum) emission levels. This has been successfully countered by the argument that to reduce the emission from 50 mg Nm^{-3} down to 25 mg Nm^{-3} consumes a great deal of additional power. Figure 6.7

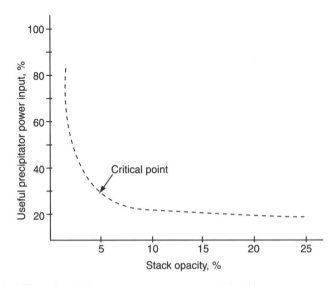

Figure 6.7   Effect of precipitator power input on opacity level

indicates that the additional power required on a large precipitator installation to reduce the emission by 50 per cent can be significant, hence the overall effect on global emissions in winning the fuel, transporting and preparing it for combustion and then having a 30 per cent electricity conversion efficiency, is significantly greater than that resulting from the slightly higher particulate emission. In any event, various companies have developed critical point control (CPC) type algorithms for controlling the power usage, which overcomes the 'set-point' approach to some extent [3]. It will be appreciated that the efficiency of any given installation, and hence performance, is controlled by the specific power usage. For plants where the gas flow is lower than the design volume or the fly ash precipitates more readily than anticipated from the firing of a higher sulphur or lower ash content fuel, then the basic design efficiency can be readily exceeded. This in turn not only results in a very low emission but a significantly increased power consumption because of the reduction in space charge effects. It will be seen from Figure 6.7, which relates stack opacity/emission to specific power input, that, under typical operating conditions, as the power is reduced the opacity initially remains fairly constant, until at a certain power level it begins to increase rapidly.

By choosing a certain opacity level as a critical control point (CP), it is possible to curtail only the excess power used by the precipitator. By continuously monitoring opacity levels this routine assures that input power levels are reduced only to the point of a defined opacity rise (as low as 0.1 per cent). A set point is not used, and the system accounts for boiler load swings and possible changes to fuel blends by searching for the optimum performance levels based upon these factors. Figures 6.8 and 6.9 illustrate a typical control operation on an existing power plant precipitator installation handling a relatively easy fly ash, indicating it is possible to control the emission satisfactorily even following a power reduction of some 50 per cent.

6.4.3.2 Energy management systems

The principle used is depicted in Figure 6.10, which shows a three-field precipitator with two bus sections per field, each section being energised by a separate TR set, which in turn is controlled by its own AVC unit.

The opacity (or extinction) meter is mounted in the stack and delivers a 4–20 mA signal to a converter, where the opacity signal can be filtered, converted to a digital signal, etc., before it is applied to the control units. This signal is a measure of the actual dust emission in the stack, based on a gravimetric measured calibration, and is compared with a set point in each control unit, which results in a certain corona power control action in order to accomplish a particular objective.

This is a simple and economic solution and must not be confused with the more expensive approach, where the control units are connected to a common communication bus, which is connected to a 'master' or plant computer, via a 'gateway' unit. This computerised central control system for precipitators is covered later in a following section.

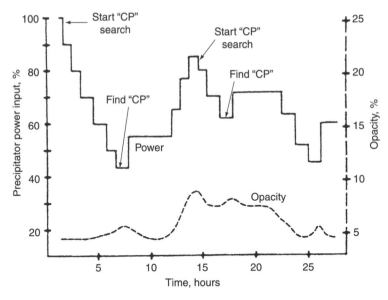

Figure 6.8   Critical control point operation

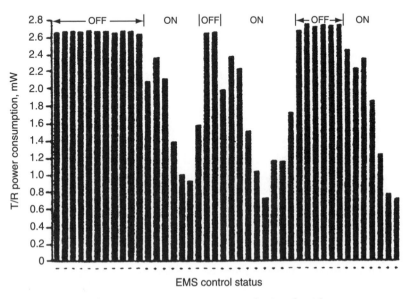

Figure 6.9   Results of power optimisation using critical point algorithm

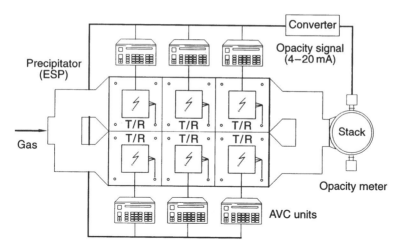

Figure 6.10   Simple energy management system (EMS) for an ESP

6.5 Basic AVC control principles

Under ideal operating conditions, the firing angle of the thyristors could be controlled manually, but in practice this is impossible. Most of the processes, which use electrostatic precipitators for particulate control, are subject to both slow and fast changes in the inlet operating conditions, e.g. the gas flow, gas temperature, gas humidity, fuel mix, raw material, etc., which can rapidly alter. In order to keep the collection efficiency as high as possible under difficult conditions, more powerful and sophisticated control units are continuously being developed and applied.

One basic AVC architecture is illustrated by the block diagram in Figure 6.11, where the precipitator current mA is used as the feedback signal, i.e. the precipitator mean current is the controlling parameter in a closed loop. In other words, the firing angle of the thyristors is varied by a proportional integral (PI) controller in such a way that the mean current follows a reference signal (or a time varying set point) as closely as possible.

The firing pulses to the thyristors are delivered by a controlled AVC output stage, providing an adequate firing signal level and electrical isolation from the a.c. line. The kV signal is also shown in the figure connected to the control unit, but in this arrangement it is mainly used in connection with spark detection and voltage recovery as explained in the next section.

Most control units are now based on microprocessors and peripheral circuitry, which offer very powerful performance because of their inherent memory and computing capabilities. Generally the programming is such that any chosen reference signal varies as a function of time, according to the control strategy.

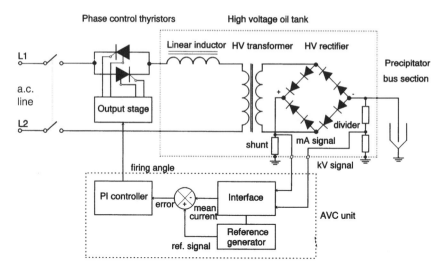

Figure 6.11   Principle of the closed loop automatic control of the precipitator current

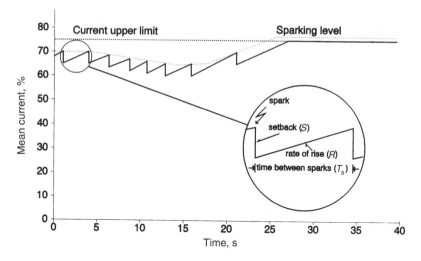

Figure 6.12   Basic control strategy utilising the precipitator mean current (courtesy FLS Miljö a/s)

An example of this basic control principle is illustrated in Figure 6.12, where, the mean precipitator current is increased linearly at a rate of rise R, until a spark occurs or an upper limit is reached. R is normally expressed in percentage min⁻¹ where 100 per cent corresponds to the rated mean secondary current of the TR set.

In this example, one assumes that the sparking level changes in the way shown, that is, it is fairly consistent at the beginning, then reduces, remaining low during a short period of time before increasing again. When a spark occurs, the current is automatically reduced by a constant setback value S, where S has an absolute value expressed as a percentage of the rated current, e.g. 5 per cent. If one assumes a constant sparking level, the spark rate SPR can be expressed as the reciprocal value of the time interval between successive sparks, T_s.

From the zoomed area in Figure 6.12, it can be seen that the rate of rise is determined by

$$R = S/T_s \text{ percentage min}^{-1} \tag{6.19}$$

and the spark rate is expressed by

$$SPR = \frac{1}{T_s} = \frac{R}{S} \text{ sparks min}^{-1} \tag{6.20}$$

Equation (6.20) indicates that a high sparking rate can be obtained with a high rate of rise R and a small setback S, the controlled variable. Conversely, a low rate of rise and a large setback produce a low spark rate. In the example shown in Figure 6.12, the rate of rise $R = 100$ per cent min^{-1} and the stepback $S = 5$ per cent; then the spark rate, at a stable sparking level, will be around 20 sparks min^{-1}. When the controlled variable reaches the upper limit the spark rate becomes zero.

The parameters R and S or S and SPR are normally programmed as settings in all control units. S is used as an absolute or relative value. The way these parameters are used and set by the specialists of the precipitator suppliers is different, each advocating their system as having the best control strategy. Often this is based on tradition, for the processes where the control units are normally used or are associated with particular characteristics of their precipitator design.

Because of the variety of control strategies used, e.g. current/spark rate, voltage hill climbing, etc., only the general principles will be reviewed. In assessing a particular control unit or design strategy it is important to recognise that the approach has been proven in practice for difficult processes, e.g. those with fast varying operating conditions, like metallurgical plants, cement kilns, etc. A further consideration is to establish that the control unit has been developed in close association with a precipitator manufacturer, i.e. by people with experience in precipitation theory and operation.

With respect to the basic control strategy as shown in Figure 6.12, in order to maintain a high corona power level during varying inlet conditions, the rate of rise R has to be large, the stepback S has to be as small as possible and the spark rate SPR has to be high. There is, however, an upper limit for the spark rate, above which the collection efficiency starts falling because of 'precipitation time' lost during voltage recovery following a flashover. Moreover, too high a spark rate may be detrimental to the life of the internal parts of the precipitator and the high voltage equipment.

One of the important objectives in a modern control unit is to obtain a fast recovery of the precipitator voltage after a spark; in this way it is possible to maximise the voltage–time integral and maintain high collection efficiency. A fast voltage recovery is only obtained if

(a) unnecessary turn-off time intervals of the control thyristors are avoided (this turn-off time is also referred to as the 'deionisation time', 'quench time', etc.) and

(b) the voltage is quickly raised to the highest possible level within a few half-cycles of the line frequency.

It is also imperative to perform this voltage recovery without a new spark arising; i.e. 'multiple sparking' should be avoided. Voltage recovery is, however, closely related to spark detection, so this aspect will be briefly examined in the following sections.

Generally, sparks are automatically classified by most controllers, into two main types according to their intensities, for example: (a) light sparking (or spitting), where the precipitator instantaneous voltage rapidly recovers to a certain level within a very short time period; and (b) severe sparking (or arcing), where the precipitator instantaneous voltage remains low for a significant period of time before recovering.

Figure 6.13 illustrates the two types of spark and the level of voltage recovery achievable by a good modern control unit. This also shows that even in the case of a severe spark, the voltage can be rapidly raised to a high level, without the utilisation of turn-off time and without the occurrence of multiple sparking.

The question of whether or not to utilise a thyristor turn off time to control sparking is one of the less understood problems in automatic voltage control techniques. There are some manufacturers of modern control units who recommend in their user manuals the use of a turn-off time in order to avoid the occurrence of arcs in the precipitator, while others do not. The alternative method to avoid such a problem arising will be explained in the next section.

A fast voltage recovery is also closely related to the method of detection used; in the past, the primary values and the precipitator current have been used, but the use of the instantaneous precipitator voltage has proven to be superior for controlling the voltage recovery. The problem of recovering the precipitator voltage within a few half cycles of the supply frequency without introducing a

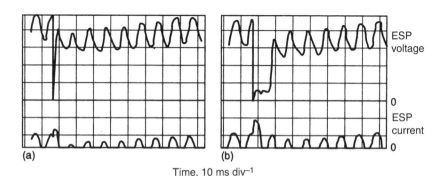

Time, 10 ms div⁻¹

Figure 6.13   Classification of sparks according to their intensity: (a) light sparking and (b) severe spark (short circuit)

specific turn off time is complex. One of the difficulties is to know how much the instantaneous voltage can be raised without the re-occurrence of sparking, i.e. the determination of the 'aimed level', and the next, to find the predetermined value of the thyristor firing angle, which will provide this 'aimed level'.

This problem and its solution are illustrated by means of the curves presented in Figure 6.14. Curve B shows the typical variation of the mean voltage as a function of the firing angle during d.c. normal operation, while curve A shows the attainable peak voltage in the first half-cycle after the spark. Experience has shown that the aimed level can be represented by the curve B, without causing multiple sparking, and at the same time producing an acceptable precipitator voltage level. Sometimes a higher aimed or target level might be used, but the probability of sparking in the recovery period would be quite high.

The problem is possibly best illustrated by the following example, where it is assumed that the control unit is firing the thyristors at 80° and the 'aimed level' for voltage recovery of 52 kV is shown by the dotted line. After a spark, the preset stepback gives an increased firing angle, a_1, which will give a voltage level determined by the intersection of curve A. If, however, the firing angle is a_2, producing a voltage level of 70 kV, this is obviously too high, compared with the 'aimed target level', and will undoubtedly result in multiple sparking. The correct firing angle is determined by the intersection of the dotted line representing the aimed level and the curve A.

At lower voltage levels, i.e. beyond the crossing of curves A and B, the problem is reversed: if the closed loop control is not opened, the firing angle will be too high and results in too low a voltage level, accompanied with a slow voltage recovery. The recommended solution to this problem would be to:

(a)   store the curve A in the memory of the control unit;

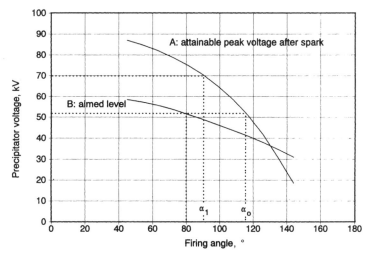

Figure 6.14   Control principle for a fast voltage recovery after breakdown (courtesy FLS Miljö als)

(b)   open the control loop in case of a spark and find the right firing angle according to the aimed level,
(c)   close the control loop and perform the required stepback.
(d)   continue with the normal control strategy after the stepback is performed.

To illustrate that a satisfactory voltage recovery can be obtained automatically, both at high and low current operation, and without using turn-off times, the oscillograms shown in Figure 6.15 are included. They speak for themselves and no further explanations need to be given.

Other important features obtained with this method of control can be seen in Figure 6.15:

(a)   At higher current operation, the first pulse current used to raise the voltage to the aimed level, and the immediate following ones, are lower than the current pulses at normal operation.
(b)   At low current operation, the current pulse used to raise the voltage is higher than at normal operation.

It can be concluded that the electrical equipment is not subject to an overload condition, in connection with a spark or arc, if the above-mentioned method is used. In this respect, it is necessary to remember that the condition for obtaining this result is the use of a suitable high short circuit reactance.

6.6 Back-corona detection and corona power control

The occurrence of back-corona in one or more precipitator sections has been normally determined by examining the corresponding V/I curve. In the past, this curve was measured as a plot of the precipitator mean current against mean voltage. The criterion used for determination of back-corona was the slope of the V/I curve; if the slope was negative (falling kV), it indicated back-corona. But in the early 1980s it was considered that this method was not sensitive enough [1], and a better indication was obtained by using a V/I curve, where the

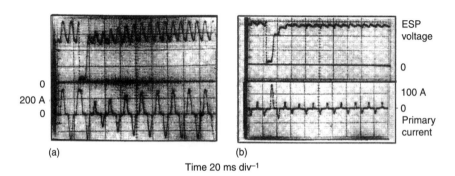

Figure 6.15   Oscillograms of current waveforms (courtesy FLS Miljö a/s)

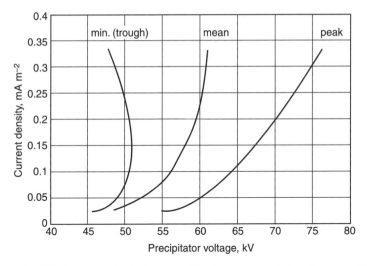

Figure 6.16 *V/I characteristics in the case of back-corona conditions (courtesy FLS Miljö a/s)*

mean current is plotted against the minimum value of the precipitator voltage, as illustrated in Figure 6.16.

This shows that the curve taken as a function of the minimum voltage changes slope at a relatively low current level, while the curve based on the mean voltage still has a positive slope. The curve based on peak voltage shows a positive slope, irrespective of the presence of back-corona. Not all voltage control units include means of back-corona detection, but those which have back-corona detection invariably use the following control principles.

(a) The slope of the *V/I* curve. This corresponds to the method described above, where the minimum voltage of the precipitator is used. This method, however, has the disadvantage that the power levels have to be systematically reduced in order to find the actual inflexion point.

(b) Voltage waveform at sparking level. This method is also based in the minimum value of the precipitator voltage, and by comparing its value before and after a spark the occurrence of back-corona can be determined. Back-corona exists if the minimum value after the spark is higher than the value before the spark.

This approach has the advantage that it does not need to reduce power levels, because sparking is used as the monitoring system. In the case of 'no sparking' conditions, a blocking period, where the thyristors are not fired, is used instead.

By observing the waveform oscillograms shown in Figure 6.13 and Figure 6.15, it can be concluded that back-corona does not occur in either of these instances.

6.7 Specific power control

6.7.1 Control by the use of intermittent energisation

In the incidence of back-corona, in order to extinguish it and maximise performance, the corona power has to be reduced. As seen in the previous section there are two means to accomplish this:

(a) to reduce the electric charge (Q_p) delivered to the precipitator, i.e. the area under a current pulse, by delaying the firing angle of the thyristors;
(b) to change to intermittent energisation sequence and find the optimal degree of intermittence.

Plant personnel have traditionally performed this task manually, but over the past decade the tendency has been to do it automatically. One approach has been to include this function into the AVC units; but it is difficult to obtain the optimal settings when only the AVC unit uses the normal instrumentation signals, as shown in Figure 6.11.

The optimisation of the degree of intermittence (D) can be performed, however, in combination with the automatic detection of back-corona. If during the detection back-corona is found, then D is increased, but if no back-corona is detected, D is reduced. The optimisation of the electric charge delivered by each current pulse, Q_p, is difficult. One method employed is to maximise the minimum value of the precipitator voltage while remaining at the optimal degree of intermittence.

A combined method also exists for the optimisation of D and Q_p [4]. Here D and Q_p are varied, so a variety of combinations of these two quantities are obtained. For each combination, a figure of merit is determined and the combination giving the best figure of merit is selected for control. An alternative method of determining the figure of merit is to compare the instantaneous voltage waveform with a reference voltage during the time interval when the corona discharge takes place, the reference voltage, for convenience, being set at the corona onset voltage V_0.

This method is depicted in Figure 6.17. The figure of merit can be determined as the integral:

$$I = \int v_p(t) - (v_p(t) - V_{ref}) dt. \tag{6.21}$$

Other methods of determining a figure of merit have been developed, but these will not be reviewed.

In order to improve the above-mentioned tasks, some units also incorporate the opacity signal in the optimisation of the corona power, as shown in Figure 6.10. However, this approach can never be 100 per cent effective, as a particular control unit does not know if there has been a change in the settings of another control unit or a variation in the operating conditions of the precipitator. Therefore, the present tendency is to place this task in a 'supervisory computer', which receives all the relevant process and stack emission signal information. With

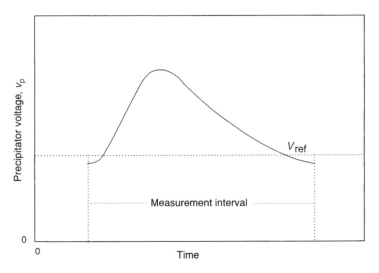

Figure 6.17   Determination of the figure of merit based on a comparison of the precipitator voltage with a reference voltage

these data the computer can then optimise the settings of the individual control units accordingly.

6.7.2 Supervisory computer control using a gateway approach

A typical solution is as depicted in Figure 6.18. The common communication bus for the control units is a traditional accepted industrial standard; however, if this is not available, the feature has to be included in the AVCs or a suitable converter has to be used. When this requirement is met, the gateway unit makes communication possible between the standard bus and various types of common PLC system. This communication problem is similar for other types of equipment used in industrial plants, which has led to the development of the required low cost communication drivers, and these gateway units can now be obtained as stock components enabling their widespread usage in the precipitation industry.

Once a connection of the AVCs via a standard bus and a gateway to the computer has been established, the supplier normally has other modules that can be connected to the standard bus to exchange data with each other as shown in Figure 6.19. The standard bus runs through the whole plant and different modules can be physically connected at the point where they are required. For example, a terminal for remote operation of the AVCs can be connected, if neither local nor control room operation is required, or, different intelligent I/O units or small PLCs can be connected via a gateway unit to the standard bus, which, because of its flexibility and modularity, makes this an attractive approach.

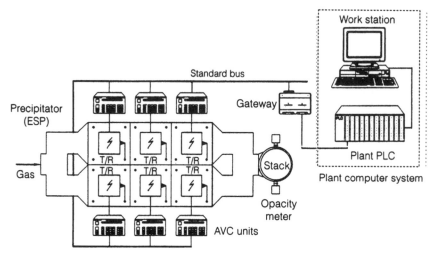

Figure 6.18   Supervisory computer control from the plant computer system via a gateway unit (courtesy FLS Miljö als)

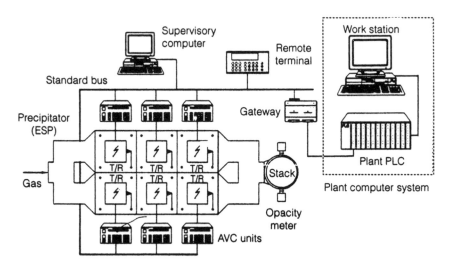

Figure 6.19   Supervisory computer control with dedicated ESP computer integrated with the plant computer system by means of a gateway unit (courtesy FLS Miljö als)

6.8 Advanced computer control functions

Among the features offered by a precipitator supervisory computer, one of the most important is that of an advanced control strategy including functions like optimisation of:

(a) the corona power,
(b) the degree of intermittence (D),
(c) the charge delivered per current pulse (Q),
(d) the spark rate and current setback, etc.,
(e) the rapping sequences, including optimisation of the off-time between rapping of the collecting plates,
(f) synchronisation between rapping of adjacent sections, programmable power-off rapping sequences, etc.,
(g) energy management and control.

A supervisory computer is able to fulfil these requirements in a satisfactory way because it is constantly receiving information about:

(1) process condition (gas temperature, flow, feed rate, O_2, SO_2, etc.),
(2) stack opacity (dust emission),
(3) currents and voltages from the individual bus sections,
(4) status of TR sets, rappers, timers, parameters, etc.

Based on this data the supervisory computer can therefore optimise the relevant settings in each automatic control unit (AVC), such as:

(a) the degree of intermittence (D),
(b) the upper current limit (Q_0),
(c) the spark rate SPR and voltage stepback S,
(d) the rapping off time, etc.

These settings are changed at regular intervals, or when required, and their positive effect assessed by analysing the opacity signal, while simultaneously checking that the process conditions have not changed by means of a trend analysis of their respective signals. These relevant settings cannot be optimised to the same degree by the AVCs operating as stand-alone units, i.e. only relying on the electrical mA and kV feedback signals.

For all these reasons, it is considered that the architecture depicted in Figure 6.19 will become more and more accepted. The control room personnel will operate the precipitator from their monitor stations in the normal way but via a gateway communication with the precipitator supervisory computer running in the background. This computer will overtake and perform more and more advanced control and monitoring functions, like precipitator event and alarm indication and handling, fault diagnosis, etc. These features, complemented with precipitator startup and shutdown automatic routines, will result in an intelligent and powerful supervisory computer control, thereby relieving plant personnel from tedious work routines and will provide cost advantages because of power savings, easier and better maintenance, higher plant availability and a lower average dust emission.

6.9 References

1 REYES, V.: 'Comparison between Traditional and Modern Automatic Controllers on Full Scale Precipitators'. Proceedings 7th EPRI/EPA Symposium on the *Transfer and Utilisation of Particulate Control Technology*, Nashville, Tennesee, USA, 1987, Session 2A, EPRI Palo Alto, Ca., USA
2 REYES, V.: 'Methods and Apparatus for Detecting Back Corona in an ESP with Ordinary and Intermittent Energisation'. US Patent No. 4,936,876, June 26th 1990
3 WEINMANN, S. F., and PARKER, K. R.: 'Meeting Emission Levels through Precipitator Upgrades'. Proceedings 8th EPRI/EPA Symposium, San Diego, Ca., USA, 1990, Session 8A; EPRI, Palo Alto, Ca., USA
4 JACOBSSON, H., and PORLE, K.: 'Method of Controlling the Supply of Conditioning Agent to an Electrostatic Precipitator'. PCT Patent Application WO 94/20218, September 15th 1994

Chapter 7

Alternative mains frequency energisation systems

Although the majority of existing operational single-stage precipitators use full wave mains rectification for energisation, there is a growing trend to adopt other means of energisation to overcome the deleterious effects of reverse ionisation when handling high resistivity dusts, and/or to optimise power consumption on large installations. These alternative systems will be reviewed in this chapter.

7.1 Two-stage precipitation

In the past, two-stage precipitation has been the preserve of air cleaning duties; however, there have been a number of recent investigations to apply two-stage precipitation, in a hybrid arrangement, to improve fly ash collection from power plant and other applications.

7.1.1 Air cleaning applications using positive energisation

For air cleaning duties, in such locations as clean rooms, hospital theatres, offices, etc., the quantity and size of particulate to be removed is very small but contained in large air volumes. Because the cleaned air will be used mainly for breathing, the system adopts positive ionisation for both particle charging and collection, since the ozone production rate considered hazardous to mankind is lower than that from negative ionisation. The use of positive energisation is in spite of negative energisation having the lower corona inception and higher breakdown voltage and hence enhanced performance for the same operating conditions.

Because of the low particulate concentrations and fine particle sizing, two-stage precipitation, comprising separate charging and precipitation zones are used, as opposed to single-stage precipitation where both charging and precipitation occurs with a single set of electrodes. Figure 7.1 illustrates a typical electrode format for two-stage precipitation.

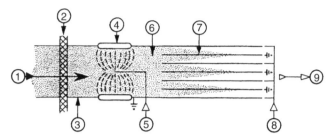

1. Air flow
2. Pre-baffle removes coarse material
3. Atmospheric smoke and larger particles carrying relatively weak charges
4. Nonlinear electrostatic forces produce powerful field of charged air molecules
5. High tension current to ioniser wire
6. Highly charged dust particles
7. Linear electrostatic field forces charged dust to collector plates
8. High tension voltage to alternate plates
9. Filtered air

Figure 7.1   Typical arrangement for two-stage precipitation

In practice, the charging section typically takes the form of a single discharge element mounted between short parallel earthed plates spaced at some 25 mm. This charging section is then followed by a pure precipitation section comprising an arrangement of alternate charged and earthed plates, typically spaced at around 12 mm. The charging section is usually energised at + 12–15 kV, while the precipitation section operates at some + 6 kV. The design of the energisation equipment depends on the manufacturers' preference, running initially from simple voltage doubler type arrangement, through electronic voltage amplifiers to high frequency d.c. derived high voltages, as used in TVs, PCs and similar commercial applications, cost being the prime concern in a competitive environment.

A typical air cleaner cell is illustrated in component parts in Figure 7.2.

With the large air volumes usually treated in practice, a large number of single cells are operated in parallel and with the low particle concentrations the cells are not rapped during operation. With the relatively high face velocities, up to 10 m s^{-1}, the discharge electrodes, usually very thin tungsten wires to withstand incipient flashover, remain fairly free of deposition. The precipitation zone plates, typically fabricated from thin sheet material, such as an aluminium alloy, are arranged as a cassette for easy removal and cleaning. The plates are usually coated with oil or a similar substance to eliminate particle re-entrainment resulting from the high face velocities. Some installations, however, employ liquor flushing/washing for *in situ* removal of any deposits, rather than removing the cassette arrangement for off-line cleaning.

Depending on the operational face velocities, removal efficiencies of up to 99.5 per cent are readily obtainable with air cleaning precipitators. Ozone production, which is typical of ionisation, is maintained at a low value because of

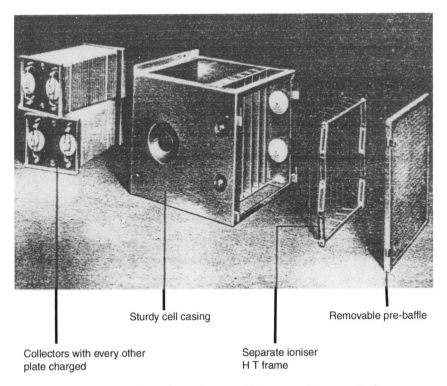

Sturdy cell casing   Removable pre-baffle

Collectors with every other
plate charged

Separate ioniser
H T frame

Figure 7.2   Photograph of a single air cleaning cell (courtesy Sturtevant Ltd)

the positive ionisation plus high gas velocities passing the discharge element, which minimises the available contact time in the reactive area of ionisation.

7.1.2 Two-stage precipitation as applied to power plant precipitators

When handling high resistivity fly ash, precipitator performance is disadvantaged by the emission of positive ions from the deposited dust layer on the collectors. This situation, as reviewed later, is exacerbated by the continuous arrival of charge. As it was known that particles escaping, usually as re-entrainment, from an upstream field in a conventional precipitator still retain a high charge ratio, a downstream d.c. non-emitting field should satisfactorily collect these pre-charged particles.

Initial investigations applying two-stage precipitation were carried out in the UK Research Laboratories of the CEGB using a large pilot plant powered by conventional negative energisation, in which the outlet field was converted to a low turbulence zone, by replacing the conventional discharge electrodes by profiled plates having low emission characteristics. The results of testing indicated that under high resistivity fly ash conditions, the performance, over a conventional discharging field, was enhanced by a 30 per cent increase in the

Deutsch effective migration velocity. These laboratory results were initially considered satisfactory as a possible means of overcoming the deleterious effect on existing installations, when faced with handling high resistivity fly ash from a reduced sulphur coal.

Full scale trials were subsequently carried out on large conventional twin flow, three-field precipitation units. One flow retained the standard discharge electrodes in the outlet field, while the other flow outlet field was converted to an all plate configuration. This arrangement enabling direct performance comparison to be made between the conventional and modified unit. The results of parametric testing, under a non-rapped outlet field situation, indicated that, when treating a normal resistivity fly ash from a 2 per cent sulphur coal, the modified flow did not significantly improve the overall collection efficiency. With high resistivity fly ash arising from <1 per cent sulphur coal, however, the overall performance was substantially increased over that obtained from the conventional precipitator. This increase in performance was attributed partly to the much higher voltage and hence stronger electric field applicable to the all plate zone, but was mainly a result of the absence of reverse ionisation characteristics. The only charge arriving on the plates is that carried by the particulates from the upstream fields, which was able to leak satisfactorily to earth, without the deposited layer developing a voltage high enough to emit positive ions.

Although the results of two sets of measurements, employing different plate configurations in the outlet field, confirmed the increased performance obtained during of the initial pilot trials, overall, the performance increase, as a Deutsch migration velocity for the whole precipitator, was only around 10 per cent. This increase was not considered viable in terms of overall cost effectiveness for retrofitting plants using lower sulphur content fuels to mitigate SO_2 emissions.

One difficulty with the approach in converting only the outlet field was that upon rapping any ash re-entrainment passed directly into the outlet flue without any means of recharging and hence recapturing the slip. This led to investigations in the UK using a hybrid combination with the all plate zone being installed as a second field, where not only is the slip from the first field higher, but with the larger particle size, the slip would carry a high charge ratio, hence, the collection efficiency of the all plate non-emitting zone should be enhanced. The improved collection efficiency being obtained with this approach is not only because the final field acts as a policeman to prevent re-entrained particles from reaching the outlet flue without being precipitated, but it was found that in the second field, the presence of both fine and coarse particles in the high field intensity resulted in significant particle agglomeration, making them easier to charge and collect.

This hybrid approach was tested on a full scale plant in the US; again, the unit converted was a three-field precipitator where the second field was modified to include profiled low emission plates in place of the conventional discharge electrodes [1]. This plant fires a coal producing an ash having an electrical resitivity above 10^{11} Ωcm and to maintain emission compliance SO_3 is injected as a matter of course. This complicated the testing procedure, since without conditioning the opacity exceeded the permitted target value. Prior to the conversion, in order to establish a base level, emission measurements were made with reducing SO_3 injection rates; from these tests the emission without injection was evaluated as around 120 mg am^{-3} at full boiler load. Tests were then carried out with the modified plant, under both conditioning and non-conditioning operation. Without conditioning, the average emission was determined at around 48 mg am^{-3}, whereas with both low and high SO_3 injection rates, the emission decreased to around 44 mg am^{-3}. These results indicating that the conversion of the second field to an all plate precipitation zone significantly reduced the impact of the relatively high fly ash resistivity. A further point was that without conditioning the emission was maintained, whereas previously, without conditioning, the opacity rose to exceed the permitted level within a very short time frame. Evaluation of the test data obtained from these tests, which covered a period of several months, was that the conversion resulted in a minimum increase of 40 per cent rising to 100 per cent in the Modified Deutsch migration velocity (ω_k).

A similar investigation has been carried out on a large three-field pilot precipitator collecting salt cake from a slipstream of a pulp and paper mill boiler. The fly ash (salt cake) from this process is mainly submicrometre in size and very cohesive; nevertheless, test data indicated with the all plate zone in the second field of the pilot that there was around a consistent 70 per cent increase in the modified Deutsch migration velocity (ω_k). This was mainly attributable to improved particle agglomeration resulting in their receiving a higher charge, together with the outlet field collecting rapping slip from the all plate zone, producing an enhanced collection efficiency [2].

Although the results of this two-stage hybrid approach have indicated positive performance enhancements, particularly when handling fine particulates, the retrofitting of existing installations (because of the need for specially profiled low emission collectors and overcoming the additional weight on the structure) can prove expensive. Consequently, if an upgrading situation arises then alternative approaches may prove significantly more cost effective.

7.2 Intermittent energisation (IE)

Whereas most operating precipitators use full wave high voltage rectification for energisation purposes, there is, particularly for larger installations, a requirement to minimise energy consumption. This can be achieved by an intermittent energisation (IE) approach, where the thyristor firing is held off for specific periods based on feedback from an opacity meter as reviewed in Chapter 6. The use of intermittent energisation in this case, where periodic flashover assists in maintaining the emission areas fully effective, overcomes the possible deterioration in electrical operation and hence performance synonymous with the earlier system of reducing the operating voltage in order to effect power savings. This IE approach can also be beneficial in the case of reverse ionisation arising with high resistivity fly ash, when the time interval between successive current pulses enables charge to dissipate through the dust layer, rather than allowing a voltage to build up on the surface.

Intermittent energisation (IE) is a system introduced with the dual purpose of saving energy and improving the collection efficiency of precipitators when handling high resistivity dusts. This energisation form is also known under other trade names like energy control [3], semi-pulse [4], Variopuls [5], etc. IE operation emerged as a cheaper alternative to the pulse energisation that had been developed and used commercially in the solution of high resistivity dust problems.

7.2.1 Basic principles of intermittent energisation

Intermittent energisation is obtained with the same electrical equipment as employed in traditional full wave power supplies. The difference resides in the automatic voltage control equipment, which has the capability of suppressing a certain number of half cycles of the primary current delivered to the transformer by the a.c. supply.

The suppression is obtained by not firing the phase control thyristors in the respective half cycles or, alternatively, by using a firing angle of 180°. The principle of operation is illustrated for a 50 Hz rectified supply by the waveforms shown in Figure 7.3. This example shows six half cycles of the line frequency and indicates the case where two out of three current pulses are suppressed. Figure 7.3(a) shows the primary current in relation to the supply voltage, the corresponding waveforms of the precipitator voltage and current are shown in Figure 7.3(b); these waveforms were obtained with a firing angle corresponding to 3 ms after the zero crossing of the line voltage.

The waveforms in Figure 7.3 show the following advantages compared with d.c. energisation for the same firing angle as in Figure 6.2:

(a)   the peak value of the precipitator voltage is higher,
(b)   the minimum (trough) value of the precipitator voltage is lower, and
(c)   owing to the suppression of two current pulses, the mean and the r.m.s values of the precipitator current are significantly reduced.

With intermittent energisation, since the current to the precipitator is discontinuous, the operation is specified in terms of the degree of intermittence, or *D*, which is derived by dividing the number of half cycles included in one energisation duty cycle by the number of current pulses in this time interval. In the example shown in Figure 7.3 the energisation cycle is three, so the degree of intermittence *D* is 3. In another example of IE, where the thyristors are fired once and then kept blocked for the next 10 half cycles, then the degree of intermittence *D* would be 11.

The degree of intermittence is also expressed by other names, for example, the 'charge ratio' δ, which is defined as the number of current pulses in one duty cycle divided by the number of half cycles included in the energisation cycle. For instance, the charge ratio, δ, in Figure 7.3 is 1/3, that is, the waveform comprises

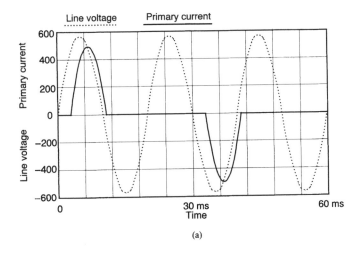

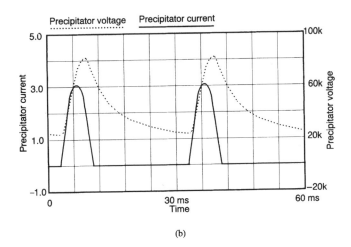

Figure 7.3   Typical waveforms with intermittent energisation

one current pulse out of every three. This means that the charge ratio (δ) is basically the reciprocal of the degree of intermittence (D).

The waveforms for IE presented in Figure 7.3 and Table 7.1 (courtesy FLS Miljö a/s), present a direct comparison of the electrical operation between intermittent energisation, with a 3 ms firing angle, in comparison with conventional d.c. energisation.

7.2.2 Comparison of IE with traditional d.c. energisation

As a consequence, the impact of IE on precipitator operating conditions compared with d.c. energisation, as depicted in Figure 7.4, can be summarised as follows:

Conventional full wave T/R operation

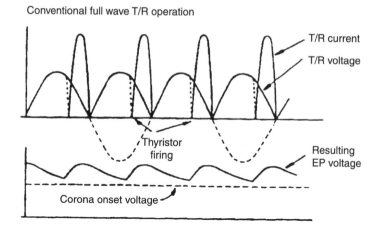

Semi-pulsed T/R operation

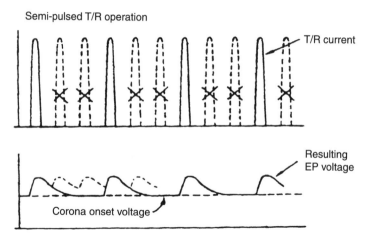

Figure 7.4   Comparison of conventional and intermittent energisation

(i)   The mean value of the precipitator voltage is lower,
(ii)   The corona power delivered to a particular bus section of a precipitator is lower, and
(iii) The power consumption of the precipitator is reduced.

The impact of these changes on operating conditions are examined in greater detail in the following sections.

7.2.2.1 Voltage levels

(a)   The peak voltage is higher because the area under the current pulse is greater. As this area corresponds to the electric charge Q_p delivered to the

Table 7.1 Current and voltages with intermittent energisation, IE, compared with values obtained with d.c. energisation

Energisation form	IE	d.c.
Firing time (ms)		3
Primary current I_p (A)	172	223
Peak precipitator current I_0 peak (mA)	3100	2350
Precipitator r.m.s. current I_0 rms (mA)	1080	1400
Mean precipitator current I_0 mean (mA)	476	1030
Peak precipitator voltage V_0 peak (kV)	82	78
Mean precipitator voltage V_0 mean (kV)	41	61
Min precipitator voltage V_0 min (kV)	24	46

precipitator section, the larger the value of Q_p, the higher the peak voltage, because the precipitator load has an inherent capacitive component.

(b) Because of the longer time interval between successive current pulses, i.e. without receiving electrical charge, the precipitator discharges towards the corona onset voltage, resulting in a lower minimum voltage value.

(c) The mean voltage is proportional to the area under the precipitator voltage in one energisation duty cycle (in this example: three half cycles of the line frequency). Because of the lower minimum voltage, the mean voltage value also reduces.

7.2.2.2 Current levels

The mean value of the precipitator current is reduced owing to the suppression of a number of current pulses, where the suppression is typically expressed as the 'degree of intermittence', D. If it is assumed the area under the current pulse is the same for both IE and d.c. energisation, and the mean current obtained with d.c. energisation is I_{dc}, then the mean current obtained with intermittent energisation I_{IE} can be expressed as

$$I_{IE} = I_{dc}/D. \tag{7.1}$$

The area under the current pulses with intermittent energisation, as shown in Figure 7.4, may sometimes be higher than that from conventional energisation by a factor k. Then, the mean current is given by

$$I_{IE} = kI_{dc}/D, \tag{7.2}$$

where k may vary between 1 and 1.5, and in the example shown in Table 7.1, k is 1.39.

In this situation the mean current is reduced by a factor k/D. (Assuming that the primary current pulse has a similar waveform in both cases, then the r.m.s. value is reduced by a factor equal to k divided by $\sqrt{D}$.)

Summarising, there are two aspects where IE can be considered inferior to d.c. energisation; these are:

(i)   the higher form factor of the primary current, and
(ii)   potential saturation of the transformer magnetic core.

The relationship between the r.m.s and mean precipitator current values in Table 7.1 clearly indicates that with IE, the form factor is higher than that obtained with d.c. energisation. This results in a corresponding higher form factor for the primary current, which is equivalent to a higher harmonic content. Also, because of the pause interval in primary current flow with intermittent energisation, the change in the magnetic induction (ΔB) is reduced, resulting in a reduced and inferior transformer core saturation. This situation can be minimised in the design using different approaches such as:

(a)   using a larger transformer core,
(b)   using a core with higher remanence and low eddy losses,
(c)   firing a smaller auxiliary current pulse before the main one,
(d)   installing a larger linear series reactor to modify the waveform.

7.2.3 Collection efficiency evaluation

Because intermittent energisation (IE) operation reduces the corona power supplied to the precipitator, and although this will result in an energy saving, it poses the question as to how this reduction in corona power influences the collection efficiency of the precipitator. The normal way to evaluate the effect of IE on performance is to measure the collection efficiency obtained with intermittent energisation and compare it with that obtained with conventional d.c. energisation. From these values it is then possible to calculate the respective effective migration velocities and hence performance factor H, which is defined as:

$$H = \omega_{kIE}/\omega_{kdc}, \tag{7.3}$$

where ω_{kIE} is the migration velocity obtained with intermittent energisation and ω_{kdc} is the migration velocity obtained with d.c. energisation.

Because the required collection surface is inversely proportional to the migration velocity for a given collection efficiency, if under IE operation the enhancement factor (H) is greater than 1, then this is the factor by which the collection surface of a d.c. energised precipitator would need to be increased to produce the same collection efficiency as that measured with intermittent energisation.

7.2.3.1 Operation under low resistivity (normal) dust conditions

The application of intermittent energisation to any precipitator because of the thyristor hold off periods invariably results in a lower precipitator mean voltage and mean current, which means a lower specific power is available for precipitation. With low resistivity fly ash, without reverse or back-ionisation being present, this reduction in specific corona power leads to a lower collection efficiency and therefore a reduced effective migration velocity.

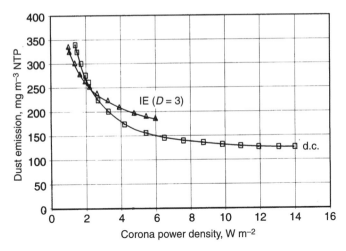

Figure 7.5   Stack emission as a function of the corona power with d.c. and IE for low resistivity fly ash conditions (courtesy FLS Miljö als)

This situation is illustrated in Figure 7.5, where dust emission is plotted as a function of the specific corona power density P_c/A for both d.c. and intermittent energisation having an intermittence factor D of 3. The precipitator relating to this figure collected fly ash from a boiler fired with a coal blend producing an ash, which did not exhibit resistivity problems. In this situation, operation under IE results in a lower performance, and therefore IE is not advantageous for this application since H is less than 1.

As indicated earlier, if an installation handling a low resistivity fly ash produces an emission lower than that specified, either because of a change in the fuels' sulphur level or ash content, the application of IE can result in a significant decrease in power while accepting a slightly lower efficiency. Reductions of up to 90 per cent in consumed power have been recorded [6, 7], without potential electrical deterioration as would result from simply decreasing the operating voltage levels.

7.2.3.2 Operation with some reverse ionisation exhibited.

In Figure 7.6, similar curves are presented for a precipitator collecting fly ash from a boiler fired with coal, which produces a fly ash resulting in a limited degree of back-ionisation. From these results it will be noted that both forms of energisation produce very similar minimum dust emissions and hence the enhancement factor is equal to 1. Therefore if a higher dust emission can be considered as acceptable (>40 mg m^{-3} NTP), then IE results in a better performance for a similar specific power consumption.

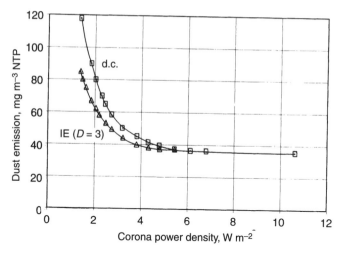

Figure 7.6   Comparison of emissions for both forms of energisation with limited reverse ionisation present (courtesy FLS Miljö als)

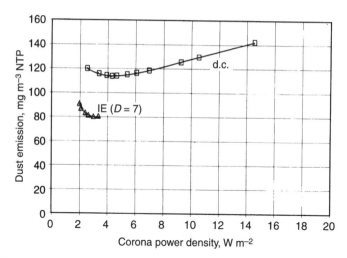

Figure 7.7   Comparison of emissions for both forms of energisation with severe reverse ionisation present (courtesy FLS Miljö als)

7.2.3.3 Operation under conditions of severe reverse ionisation.

In Figure 7.7, dust emissions are again plotted as a function of the specific power input for both d.c. and intermittent energisation having an intermittence of 7 for fly ash arising from a boiler firing a low sulphur coal, which resulted in a fly ash producing severe back-corona in the precipitator. In this instance it can be seen that operation with IE is clearly better than d.c. energisation, since it results in a lower dust emission and hence an enhanced performance factor $H > 1$ together with a lower specific power consumption.

From these data, operation under intermittent energisation in general terms can be summarised as follows.

- Intermittent energisation always results in a lower power consumption, but it does not always produce the highest collection efficiency.
- The performance with intermittent energisation is very closely related to the resistivity of the collected dust.
- With low resistivity fly ash, the performance with IE is inferior to d.c. energisation, since it results in a lower precipitator efficiency.
- For medium resistivity fly ashes, operation under IE is as good as d.c. energisation and sometimes better.
- In the case of high resistivity fly ashes, operation under IE is superior to d.c. energisation.

7.3 Pulse energisation

Another approach, initially developed as an alternative to chemical flue gas conditioning for improving the performance precipitators handling high resistivity fly ash, is pulse energisation. In this approach, short duration high voltage pulses are intermittently superimposed on a reduced d.c. precipitation voltage. The pulse produces an intense cloud of ions for particle charging and the d.c. for precipitating the charged particles, the time interval between the pulses enabling the charge to leak to earth rather than creating back-ionisation.

Historically, the first full scale pulsing tests on precipitators were carried out by White in the late 1940s [8] using a modified mechanical rectifier as a high voltage switching device. The commercial application of pulse energisation systems was hampered mainly by the lack of reliable high voltage switching equipment and by a very high active power consumption. With the advent of high frequency switching thyristors in the 1970s and refinement of control systems, companies in the USA, Japan and Europe developed specific pulsing systems and were able to demonstrate full scale tests in the late 1970s [9]. A fuller and more complete history of the use of pulse chargers can be found in Hall [10].

The advent of commercial pulse energisation systems in the early 1980s was one of the major technical developments in the energisation of electrostatic precipitators. In fact, pulse energisation was possibly the most significant change in precipitation technology since Cottrell's development of the traditional power supply based on the high voltage transformer/mechanical rectifier combination [11].

7.3.1 Introduction

Energisation using a pulsing technique was specifically developed to improve the collection of 'difficult' high resistivity particles, which results in reverse ionisation in terms of precipitator operation. The energisation basically consists of applying a very short duration high voltage pulse superimposed on a reduced 'base voltage'. The pulse systems developed for a single-stage precipitator normally operate in one of the two following pulse widths – a 1 μs or a 100 μs range; however, only the 100 μs type of pulser will be reviewed in the following sections.

The high voltage pulses are repeated at a certain frequency in the range of 1–400 pulses per second (pps). A typical waveform of the applied voltage is depicted in Figure 7.8, for a frequency of 100 pps, which also includes for comparison purposes the voltage waveform for traditional d.c. energisation. In this, the differences are quite clear in that, with pulse energisation the narrow pulse voltages have a high amplitude. With the 100 μs pulse width the peak voltage applied to the precipitator can be much higher than with d.c. energisation, without running into breakdown problems. The peak voltage amplitude is the sum of the base plus the pulse voltage, the base voltage being normally maintained close to the corona onset voltage during pulse operation.

7.3.2 Electrical configurations

As indicated in Figure 7.8, a pulse system has to produce a narrow high voltage pulse which is then superimposed onto a base voltage; consequently two supply units are required in most designs, one the pulse generating circuit and the other for the base voltage power supply. Moreover, the pulse amplitude, the base voltage and the pulse repetition frequency have to be varied according to a certain programmable control strategy requiring a special control unit to perform this function. Various manufacturers use different approaches in the design, operation and construction of their systems, mainly as a result of patent protection, so arrangements may vary from those depicted.

In principle, there are two main design architectures for single pulse opera-

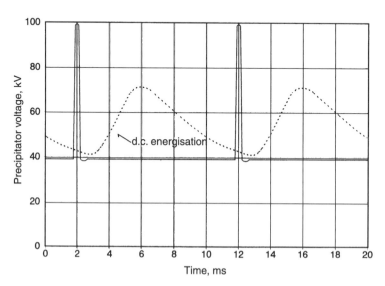

Figure 7.8   Typical voltage waveform comparison obtained with pulse and d.c. energisation

tion; one based on switching at low potential as seen in Figure 7.9 and one based on switching at high potential [12] as seen in Figure 7.10. The first type normally uses two high voltage tanks (one for the pulse generator and one for the base voltage) and separate cabinets for the automatic voltage and power control devices. The second type normally uses one control cabinet and one high voltage tank resembling a traditional power supply. Both use a series LC resonant circuit, where the precipitator, represented by its capacitance, is one of the circuit components. The high voltage initiation switch, typically a large power thyristor, has an anti-parallel breakover diode which serves to reduce power. Another approach, instead of supplying a single pulse, uses a burst of short duration pulses, derived from an inductor/capacitor discharge in the secondary circuit and is termed a multipulse system. Except for the generation of multipulses the operation and performance is very similar to the single pulse system.

7.3.2.1 *Pulsing system using a pulse transformer*

The basic circuitry is depicted in Figure 7.9. Here the base voltage is produced by a separate power supply represented by V_{dc}, while the pulse generating circuit includes a voltage power supply V_{PS} plus a series oscillating circuit, consisting of the storage capacitor C_S, the coupling capacitor C_C, the precipitator capacitance C_F and an inductance L_S.

During operation but before the generation of the pulse, C_S is charged to a voltage $-V_{PS}$ and the precipitator is charged to $-V_{dc}$. When the thyristor T is fired, discharging the capacitance, oscillation is initiated and the resultant voltage through the main precipitation circuit has the pulse waveform shown in Figure 7.8.

During the positive half cycle the current passes through the thyristor T and this is turned off around the zero crossing when the current falls below the holding level. Because of the energy stored in the inductance L_S, the breakover diode D is forced to conduct and the current circulates in the opposite direction until it becomes zero. The current remains at zero until the thyristor is again fired after a preset time period, corresponding to the pulse repetition frequency. Energy recovery, which is fundamental for the commercial utilisation of pulse energisation, occurs during the negative half cycle of the pulse, where the energy

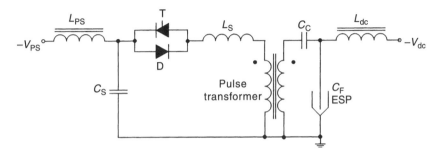

Figure 7.9   General circuit diagram of pulse transformer system (courtesy FLS Miljö als)

delivered to C_S but not used for corona generation is returned to C_F and stored during the interval between pulses but is then available for the generation of the next pulse. The coupling capacitor C_C avoids a short circuit of the power supply V_{dc} by isolating the secondary winding of the pulse transformer from the base supply. The pulse transformer approach allows switching at a low potential, but has the disadvantage of a higher price, weight and volume.

7.3.2.2 Pulsing system without a pulse transformer

This system is depicted in Figure 7.10; here the base voltage is again produced by a separate power supply represented by V_{dc}. The pulse generating circuit includes a power supply delivering V_{PS} and a series oscillating circuit consisting of a storage capacitor C_S, an inductance L_S and the precipitator capacitance C_F. In this configuration, the semiconductor switch is placed on the high voltage side and comprises a large number of series thyristors [6].

Before the generation of a pulse, C_S is charged to the voltage $(V_{PS} + V_{dc})$ and the precipitator is charged to $-V_{dc}$. When the thyristors T are fired, oscillation is initiated and the pulse voltage has the waveform as shown in Figure 7.8.

7.3.2.3 Multi-pulse operation

The circuit arrangement for this approach is illustrated in Figure 7.11. In this it will be noted that this system uses only one transformer rectifier TR set for both the base voltage and pulsing voltage supplies, the base voltage being derived by means of a bleed resistor in parallel with the main thyristor pulsing switch. This thyristor chain also has anti-parallel diodes for energy recovery, which is similar to other pulser units and is critical in terms of energy utilisation.

In operation, the capacitor C_F is negatively charged and at a certain point the thyristor is triggered and oscillation arises with the inductance capacitance combination, the pulse passing through the thyristors and appears across the precipitator, producing a waveform as indicated in Figure 7.12. The length of

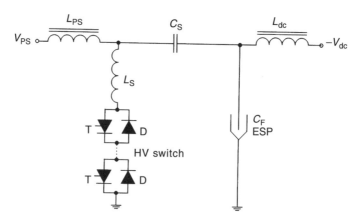

Figure 7.10   Circuit arrangement without a pulse transformer

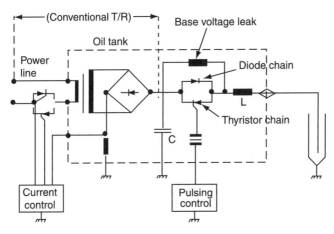

Figure 7.11   Circuit arrangement of multipulse approach (courtesy Alstom Power)

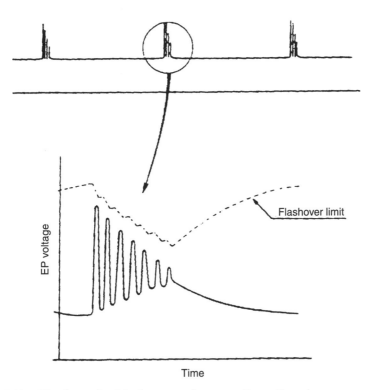

Figure 7.12   Waveform of multipulse system (courtesy Alstom Power)

the burst of pulses is controlled by the thyristor operation. Typical values for such a circuit arrangement are as follows.

Pulse amplitude	40 kV or higher.
Pulse width	90 μs
Pulses per burst	Up to 8
Pulse frequency	Up to 300 Hz

Except for the voltage waveform exhibiting a burst of pulses, the operation of the system is basically similar to single pulse operation.

7.3.3 *Electrical operation with pulse charger*

Assuming an ideal pulse transformer, no losses in the circuit and a very large storage capacitor, the pulse current can be expressed by

$$I_p(t) = I_p \sin(\theta_0 t), \tag{7.4}$$

where I_p is the peak value of the current pulse and θ_0 is the angular frequency of the oscillation, where the pulse width T_0 is normally expressed as a function of θ_0.

The amplitude of the pulse precipitator voltage V_p can then be given by

$$V_p(t) = \frac{I}{C_F} \int I_p(t)dt = \tfrac{1}{2} V_p(1 - \cos\theta t) \tag{7.5}$$

The quantities I_p, V_p, θ_0 and T_0 can be readily derived using the following equations:

$$I_p = V'_{PS}/[\sqrt{} (L'_S/C_{eq})] \tag{7.6}$$

$$V_p = 2V'_{PS} C_{eq}/C_F \tag{7.7}$$

$$\omega_0 = 1/\sqrt{(L'_S C_{eq})} \tag{7.8}$$

and

$$T_0 = 2\sqrt{(L'_S C_{eq})} \tag{7.9}$$

where

$$V'_{PS} = n \, V_{PS} \ (n \text{ pulse transformer turns ratio}) \tag{7.10}$$

hence

$$L'_S = n^2 L_S \tag{7.11}$$

and

$$C_{eq} = C_F \, C_C/[C_F + C_C]. \tag{7.12}$$

Assuming the above, the idealised pulse voltage and current would have a waveform as depicted in Figure 7.13.

In reality, the pulse transformer is not ideal and the following illustrates a

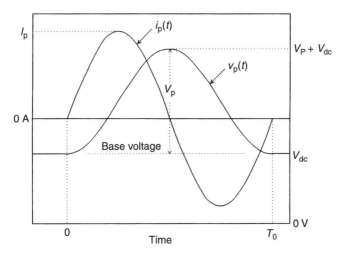

Figure 7.13   Idealised pulse voltage and current waveforms

typical practical worked example of a pulsing system using a pulse transformer.

Using typical values encountered in a practical circuit, e.g. $V_{PS} = 4$ kV; n (the turns ratio) = 10; $C_C = 500$ nF; $L_S = 25$ μH; and $C_F = 100$ nF, the amplitude of the pulse current, the amplitude of the pulse voltage and the pulse width have been calculated, from Equations (7.10), (7.11) and (7.12), as follows:

$$V'_{PS} = 10 \times 4 = 40 \text{ kV},$$

$$L'_S = 10^2 \times 25 \times 10^{-6} = 2.5 \text{ mH, and}$$

$$C_{eq} = 100 \times 500/(100 + 500) = 83.3 \text{ nF.}$$

Hence from Equations (7.6), (7.7) and (7.8):

$$I_p = 40 \times 10^3/\sqrt{(2.5 \times 10^{-3}/83.3 \times 10^{-9})} = 231 \text{ A,}$$

$$V_p = 2 \times 40 \times 83.3/100 = 66.6 \text{ kV, and}$$

$$T_0 = 2\sqrt{(2.5 \times 10^{-3} \times 83.3 \times 10^{-9})} = 91 \text{ μs.}$$

Although the above calculation specifically applies to a pulse transformer system, the general physics of the approach applies to all forms of pulser.

7.4 Major operational aspects of pulse energisation

During normal operation, the base voltage V_{dc} is maintained about the corona onset level, and the pulse amplitude and the pulse frequency are varied according to a certain preset control strategy dependent on the fly ash resistivity and electrical operating conditions. The results of using pulse energisation in a precipitator can be better expressed by the V/I characteristics of the respective bus section, as shown in Figure 7.14 [13]. Here, the specific current density, defined as the precipitator mean current divided by the collection area, is plotted as a function of the pulse amplitude for a constant base voltage V_{dc} and three different pulse repetition frequencies.

This typical family of curves illustrates the following features of pulse energisation:

- the precipitator current can be varied by changing the pulse frequency while the precipitator peak voltage $(V_{dc} + V_p)$ is kept constant,
- the precipitator peak voltage is higher than the value for conventional energisation as a result of the very short duration of the pulses,
- the slope of the *V/I* curves is rather flat, being typical for high resistivity dust conditions.

These features of pulse energisation result in the following operational advantages.

7.4.1 Current control capabilities

The *V/I* characteristic (Figure 7.14) shows that the precipitator current I_p can be controlled independently of the precipitator voltage by varying the pulse repetition frequency. This allows the current to be reduced towards the onset of back-corona, without decreasing the precipitator voltage, i.e. the precipitator can operate with a low current and high precipitator voltage. This results in a more stable and suitable electrical energisation system for high resistivity dust compared to traditional d.c. energisation, where current control cannot be achieved without reducing precipitator voltages and hence performance.

The dense ionic space charge initially produced by the pulse shields the discharge electrode and reduces the electrical field strength at its surface, which

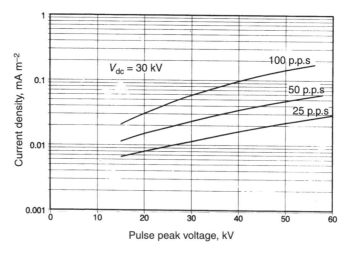

Figure 7.14   Typical V/I characteristics obtained with pulse energisation (courtesy FLS Miljö a/s)

results in the suppression or limitation of the corona discharge during the remainder of the pulse period; this allows time for any collected charge to dissipate, thereby preventing voltage build up leading to positive ion emission.

7.4.2 Current distribution on collectors

With d.c. energisation, the corona discharge tends to be localised at discrete positions on the discharge electrode and consequently the current distribution on the collectors is far from uniform. With the application of pulsing, the narrow high amplitude pulses superimposed on a base voltage around the corona onset voltage is such that the peak voltage significantly exceeds the corona onset level, which as indicated, produces an intense corona discharge and a correspondingly dense ionic space charge around the electrode. A discharge electrode with very spotty corona under traditional d.c. energisation can literally be made to glow under pulse energisation [11,14].

Investigations using d.c. energisation have shown that the collector current density at the beginning of any precipitator section is very low and increases along the section in the direction of the gas. With pulse energisation, however, the current density along the precipitator section is considerably more uniform. A good current distribution on the collecting plates is important in order to avoid the initiation of back-corona due to localised spots of high current density. This improvement in current distribution has been confirmed by measurements on a laboratory single duct precipitator and also in the field using larger pilot precipitators [15].

7.4.3 Electrical field strength in the inter-electrode area

With d.c. energisation, free electrons are constantly being generated, producing an ionic space charge density and a field strength that, in principle, does not vary with time, whereas with pulse energisation, the base voltage is kept just below the corona onset level, and free electrons and negative ions are only generated during the actual pulse period. During the pulsing, the ionic space charge rapidly crosses the inter-electrode area because of the high voltage field, while during the time interval between pulses, the space charge migrates towards the collecting electrode driven only by the base voltage. As a consequence, the space charge and the field strength vary with time at each point in the inter-electrode space.

Measurements on a tubular laboratory precipitator have produced the results shown in Figure 7.15, where the relative field strength is plotted as a function of time. This shows that, following the pulse, the electrical field is determined by the d.c. voltage and the moving space charge, its strength decreasing until the actual front reaches the collecting electrode.

7.4.4 Particle charging

For particles greater than 1 μm in diameter, collision charging is the predomin-
ant mechanism, and the saturation charge is determined by the maximum field
strength created by the ionic space charge. With d.c. energisation, a particle at a
certain position is surrounded by the ionic space charge, and its saturation
charge depends on the electric field at that position. With pulse energisation,
however, particle charging occurs only when the space charge passes the particle
and its saturation charge is determined by the maximum field strength during
the passage of that space charge. Because the maximum field strength with pulse
energisation is much higher than for d.c. energisation, as indicated in Figure
7.15, this approach provides enhanced particle charging. Measurements have
shown that the best results are obtained with a high pulse amplitude, which
results in a higher ionic space charge density and field situation.

7.4.5 Power consumption

A precipitator section can be represented by a capacitance in parallel with a
current generator accounting for the electronic, ionic and dust space charge
currents. Each pulse has to raise the voltage across the capacitance C_F, from the
base voltage V_{dc} to the peak voltage ($V_{dc} + V_P$), which involves a considerable
amount of energy usage. Supposing that in Figure 7.9 the coupling capacitor C_C
is much larger than C_F, the energy supplied by the pulsing system for charging
C_F is

$$W_C = \tfrac{1}{2} C_F V_P^2. \tag{7.13}$$

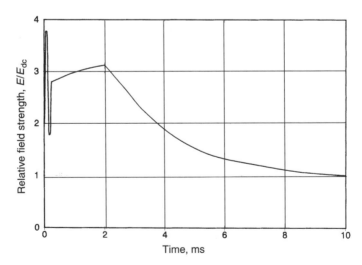

*Figure 7.15   Electric field as a function of time, following the initiation of the high voltage
pulse (courtesy FLS Miljö a/s)*

Using typical values of $V_P = 60$ kV and $C_F = 100$ nF, the energy dissipated W_C is therefore 180 J. If this has to be repeated 200 times per second, a large amount of power (36 kW) would be consumed. As the energy necessary for the corona generation is very small compared with the energy needed to charge C_F, the power consumption becomes excessive unless the pulse system includes means for energy recovery. The pulsing systems shown in Figures 7.9 and 7.10 include this energy saving feature in the form of anti-parallel diodes located in the series oscillating circuit. Here, when the voltage across C_F is at its maximum, the pulse current is zero and when the flow reverses in the negative half cycle the surplus energy from C_F is passed back into the storage capacity C_S.

The power consumed by a precipitator section energised by a pulse system with energy recovery [13] can be expressed by

$$P = I_p V_{dc} + c \times I_p V_P, \qquad (7.14)$$

where the constant c has been found experimentally to be approximately equal to 0.5.

7.4.6 Worked example of energy recovery

If it is assumed that a pulse system is operating at $I_p = 0.1$ mA m^{-2}, $V_p = 60$ kV, $V_{dc} = 40$ kV and energises a 3000 m^2 area bus section, the precipitator mean current is therefore $0.1 \times 3000 = 300$ mA, and if we apply these values to Equation (7.14) this gives the power consumed as

$$P = 0.3 \times 40 + 0.5 \times 0.3 \times 60 = 12 + 9 = 21 \text{ kW.}$$

This consumption corresponds to a power density of 7 W m^{-2}, which is a typical value for medium resistivity dusts. For high resistivity dusts, the required power density is much lower, and by operating with a low pulse repetition frequency (2–20 p.p.s.) the precipitator mean current and total power is therefore correspondingly reduced.

7.5 Collection efficiency

Similar to the case with intermittent energisation, the improvement in the precipitator performance is closely related to the resistivity of the collected dust. This improvement is normally expressed by the enhancement factor $H = \omega_{kP}/\omega_{kdc}$, where ω_{kP} is the migration velocity obtained with pulse energisation and ω_{kdc} is the migration velocity obtained with d.c. energisation.

The enhancement factor H obtained with very high resistivity dusts reported by American [14], Japanese [15] and European [16] companies increases to around 2. The enhancement factor H as a function of the dust resistivity can be expressed by the curve shown in Figure 7.16. The comparison with d.c. energisation indicates that, at low resistivity levels, both energisation forms produce the same result, but at high resistivity levels, pulse energisation is significantly better.

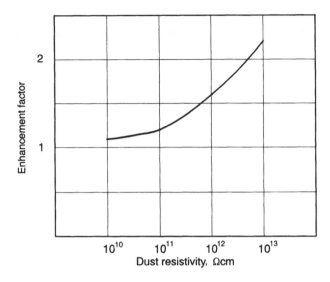

Figure 7.16   Enhancement factor obtained with pulse energisation compared to d.c. energisation as a function of the dust resistivity (courtesy FLS Miljö a/s)

Another way of expressing these results is by saying that the precipitator performance with pulse energisation does not decrease as much for increasing dust resistivities [17].

An average comparison between pulse and intermittent energisation at high resistivity levels, expressed as an average enhancement factor, gives a typical value of 1.2–1.5, favourable for pulse energisation.

7.6 Typical applications using pulse energisation

As previously mentioned, pulse energisation is used in the collection of very high resistivity dust, and because the enhancement factor is significant it can compensate for the higher price of the pulse systems compared to a much larger precipitator installation.

Typical applications resulting in enhanced precipitator performance levels with pulse energisation are:

- four-stage preheater cement kilns (without or with insufficient water conditioning of the kiln gases) [18, 19, 20];
- coal fired power and steam generating boilers [11, 16, 18, 19, 20, 21, 22, 23];
- limestone, dolomite and magnesite kilns [18];
- sinter strands for iron ore agglomeration [18, 24, 25];
- pulp and paper [19, 20];
- glass furnaces [20].

The application of pulse energisation is not restricted to new precipitators and can be a very effective solution for improving the performance of existing precipitators having resistivity problems. In this case, it is important to bear in mind that the mechanical condition of the precipitator has to be good. This means effective rapping of the discharge and collecting electrodes, good gas distribution, good electrode alignment, etc., otherwise the expected enhancement factor will not be obtained.

In practice it has been found that if a precipitator installation is collecting a high resistivity fly ash contaminated by a high carbon carryover, then since effective precipitation of the carbon particles requires a high current density, this directly conflicts with the need for a low current density to eliminate backcorona arising from the high resistivity ash. Hence the result of pulsing under this situation is a higher stack emission because of carbon slippage; therefore, a full analysis of the fly ash is necessary prior to making the decision to install pulsers.

Summarising, pulse energisation is normally ideal for precipitators collecting high resistivity dusts. The improved precipitator performance in the collection of medium and high resistivity dust, compared with traditional d.c. energisation and IE, results from the combined effect of (a) better particle charging, (b) higher collecting field strength, (c) better current distribution and (d) improved current control capability.

The enhancement factor *H*, obtained with pulse energisation, is mainly the result of the application of a high pulse amplitude creating an intense ionic cloud and improved collector current distribution. Therefore, the mechanical condition of the precipitator has to be good for the collection of fly ash and the concentration of low resistivity particles, e.g. unburnt coal, has to be low; otherwise the expected performance enhancement will not be achieved.

7.7 References

1   WETJEN, D., *et al.*: 'Enhanced Fine Particle Control by Agglomeration'. Proceedings of Power-Gen Conference, 1998, Orlando, Fl., USA
2   FELDMAN, P. C., *et al.*: 'Recent Experiences in Controlling Fine Particles in ESP'. Proceedings of the 6th International Conference on Electrostatic Precipitation, Budapest, Hungary, 1996, pp. 312–24
3   RUSSELL-JONES, A.: 'A Total Energy Management System for Electrostatic Precipitators'. Proceedings of the 3rd International Conference on *Electrostatic Precipitation*, Abano, Padova University, Italy, 1987, pp. 185–96
4   PORLE, K.: 'On Back Corona in Precipitators and Suppressing it Using Different Energisation Methods'. Proceedings of the 3rd International Conference on *Electrostatic Precipitation*, Abano, Padova University, Italy, pp. 169–84
5   NEULINGER, F. B.: 'Even Less Energy does the Job: Results of Variopuls Operation in EP Plants'. Proceedings of the 3rd International

Conference on *Electrostatic Precipitation*, Abano, Padova University, Italy, pp. 159–68

6 DUBARD, J. L., *et al.*: 'Evaluation of ESP Intermittent Energisation'. Proceedings of the 3rd International Conference on Electrostatic Precipitation, Abano, Padova University, Italy, 1987, pp. 151–8

7 SCHMOCH, M.: 'Methods to Reduce Energy Consumption of an ESP'. Proceedings of the 8th International Conference on *Electrostatic Precipitation*, Birmingham, Al, USA, 2001, Paper Session C2-2

8 WHITE, H.: 'Industrial Electrostatic Precipitation' (Addison Wesley Inc., New York, 1963, Chapter 7, Section 6, p. 232)

9 PETERSEN, H. H. et al. Precipitator Energisation Utilizing and Energy Conserving Pulse Generator. 2nd Symposium on the Transfer and Utilization of Particulate Control Technology, Denver, USA, 1979 July, EPA Vol. II, pp. 352–368.

10 HALL, H. J.: 'History of Pulse Energisation in Electrostatic Precipitation', *Jnl. Electrostatics*, 1990, **25**, pp. 1–22

11 COTTRELL, F. G.: US Patent No 895,729, 1908

12 TERAI, H., *et al.*: 'Pulse Energisation of Electrostatic Precipitators'. Proceedings of the 3rd International Conference on *Electrostatic Precipitation*, Abano, Padova University, Italy, 1987, pp. 557–69

13 LAUSEN, P., *et al.*: 'Theory and Application of Pulse Energisation'. Proceedings 1st International Conference on *Electrostatic Precipitation*, Monterey, USA, 1981, pp. 51–53

14 FELDMAN, P., *et al.*: 'Operating Results from the First Commercial Pulse Energisation System to Enhance Electrostatic Precipitator Performance'. Proceedings of the American Power Conference, 1981, Chicago, Il., USA

15 FUJISHIMA, H., *et al.*: 'Applications of an Electrostatic Precipitator with Pulse Energisation System'. 7th International Conference on *Electrostatic Precipitation*, Beijing. China, 1990 pp. 419–30; International Academic Publishers, Beijing, 1991

16 SCHIOETH, M.: 'Five Years Experience with Pulse Energisation on Power Plants Burning a Wide Range of Fuels'. Proceedings of the 3rd International Conference on *Electrostatic Precipitation*, Abano, Padova University, Italy, 1987, pp. 197–207

17 PETERSEN, H., *et al.*: 'Precipitator Energisation Utilizing and Energy Conserving Pulse Generator'. Proceedings 2nd Symposium on the *Transfer and Utilisation of Particulate Control Technology*, Denver, Co., USA, 1979; EPA, Vol. II, pp. 352–68

18 MAURITSEN, C., *et al.*: 'Experience with Pulses Energisation of Precipitators for a wide range of Process and Operating Conditions'. Proceedings of the 3rd International Conference on *Electrostatic Precipitation*, Abano, Padova University, Italy, 1987, pp 543–56

19 PORLE K. et al. Modern Electrode Geometries and Voltage Waveforms Minimize the Required SCAs. Proceedings of the 8th Symposium on the Transfer and Utilisation of Particulate Control Technology. San Diego. USA. March 1990.

20 CAPUTO, A.C.: 'Economical Comparison of Conventional and Pulsed ESP in Industrial Applications'. Proceedings of the 6th International

Conference on *Electrostatic Precipitation*, Budapest, Hungary, 1996, pp 215–20

21 CRISTESCU, D., *et al.*: 'Romanian Technologies for the Utilisation the Pulses in Electrostatic Precipitation in Energetics and Cement Industry'. Proceedings of the 6th International Conference on *Electrostatic Precipitation*, Budapest, Hungary, 1996, pp 262–73

22 YOSHIKAZU, N., *et al.*: 'Pulse Energisation for Fly Ash from Fluidized Bed Combustion'. Proceedings of the 10th Particulate Control Symposium and 5th ICESP Conference, Washington, DC, USA, 1993, Paper 19

23 YAMAMURA, N., *et al.*: 'Operating Experience of a Pulse ESP at a Modern 500 MW Coal Fired Power Plant in Japan'. Proceedings of the 6th International Conference on *Electrostatic Precipitation*, Budapest, Hungary, 1996, pp. 197–202

24 ELHOLM, P., *et al.*: 'Enhancing Sinter Strand ESP Performance Using Pulse Energisation'. Proceedings of the 6th International Conference on *Electrostatic Precipitation*, Budapest, Hungary, 1996, pp. 396–402

25 ELHOLM, P., *et al.*: 'Pulse Energisation Solving Sinter Strand ESP Problems'. Proceedings of the 8th International Conference on *Electrostatic Precipitation*, Birmingham, Al., USA, 2001, Paper Session B1-2

Chapter 8

High frequency energisation systems

8.1 Introduction

There have been few departures in the past from the use of traditional mains frequency rectified supplies for precipitator energisation, owing, in part, to the lack of availability of suitable high frequency high voltage transformers and high speed switching power thyristors; consequently the conventional 50/60 Hz design has dominated the precipitation market. The circuit architecture of the conventional design, whilst being robust and simple has, however, a number of technical drawbacks, which can be summarised as follows.

(a) The regulator output current may be discontinuous or non-sinusoidal, depending on the thyristor firing angle and the current demanded by the precipitator. As a consequence the input power factor is very poor, with a high harmonic distortion in the mains supply.

(b) As the power supply operates at mains frequency, any required step change in precipitator operation can be met only through a sluggish transition to the new working level.

(c) For optimum performance, i.e. maximum collection efficiency, it is import-ant electrically to operate the precipitator as near to its breakdown/flashover voltage as possible. This condition provides the highest voltage and maximum field strength, so that performance, which is theoretically proportional to the field strength squared, is maximised. Unfortunately, even on well-regulated and operated processes, changes in the inlet condi-tions invariably lead to flashover and hence discharging of the capacitive component of the precipitator, such that, during this time, the performance is compromised.

(d) Under certain operating conditions, as flashover can occur up to 100 times per minute, the supply must be capable of being interrupted rapidly to ensure that the flashover does not lead to an arc. With the single phase a.c. regulator this is achieved by interrupting the current for at least a full cycle of the mains supply, which can mean that the precipitator is dead for 20 ms up to 100 times per minute or more. In addition, there will be a significant time delay when the precipitator is being recharged (around 100 ms or more) on resuming the input voltage supply.

(e) Because the transformer is operating at mains frequency, its physical dimensions and weight will be high, depending on the power rating. A typical 100 kVA unit will have dimensions around 600 × 800 × 1200 mm and weigh 1500 kg or more. The transformer will almost certainly be oil insulated, which will require bund walls for containment, making civil engineering costs high. Additionally, the high reactance of the transformer core will result in the mains supply having a low overall efficiency.

Although various attempts were made to address phase balancing and poor power factor, etc., by using three phase rectification approaches, these initially experienced severe arcing that could be overcome only by completely switching off the supply. This compromised the electrical operating conditions and hence precipitator performance such that the three phase approach was not actively exploited because of the then lack of suitable switching devices for control purposes.

In the early 1990s, Siemens addressed the problem of arcing from a three phase system and developed an early form of switch mode power supply (SMPS). In this design the three phase bridge was composed of a series of diodes and thyristors, which were used as the power control element as indicated in Figure 8.1 [1]. The resulting controlled d.c. output was then chopped at a frequency of some 5 kHz to supply a fairly standard transformer. Consequently the unit was very heavy and costly: a 1500 mA unit weighing some 2 t. Although heavy, the basic design was capable of being extended to a current rating of 2000 mA. While being reasonably successful, lighter and cheaper designs of SMPS systems later rapidly supplanted commercial usage of this early low frequency system for precipitator energisation.

8.2 Development of switch mode power supply (SMPS) systems

8.2.1 Expected operational improvements

The improvements that can be obtained from using a high frequency switched mode power supply can be summarised as follows.

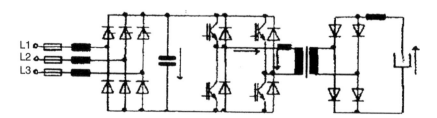

Figure 8.1   Standard three phase rectification system (SMPS)

(a) High frequency switching operation, compared with the present 10 ms periodicity, will allow a much more precise control over the precipitator operating parameters, such as operating voltage, corona discharge and flashover rates. Because the power supply will no longer be dependent on the mains supply frequency, the response of the system will quickly minimise the adverse effects of flashover, such as the short circuit duration and precipitator recharging time. Hence there will be an increase in both the precipitator collection performance and power supply efficiency.

(b) High frequency switching will allow a significant reduction in the size and weight of the high voltage transformer because flux excursions within the core will be minimised. The size may be further decreased by control of the switching voltage on the primary side of the transformer. This reduction in size and weight will lead to a more compact design, which will minimise the installation and maintenance costs.

(c) The adoption of high frequency switching will allow precise control of the electrical operation and it may well be possible to operate the precipitator much closer to its optimum working point, i.e. nearer to breakdown, with a very low incidence of flashover.

8.2.2 System requirements

A block diagram of such a system is shown in Figure 8.2. The incoming three phase supply is initially rectified, the resultant d.c. feeding a power inverter, operating at a frequency of some 20–50 kHz. This d.c. link can be modulated, or 'chopped', by using semiconductor power switches to any desired amplitude, before feeding the ferrite cored HT transformer. The transformer output, which after rectification supplies the precipitator, is controlled by varying the on/off times of the power switches using a feedback system, which continuously monitors the output voltage and current levels to achieve optimum conditions on the precipitator.

The operational problems of interrupting the power supply in order to quench an arc/flashover are considerably reduced by operation at high frequency, since at 50 kHz, the minimum switching time is 0.02 ms, compared with 10 ms at 50 Hz, and the recovery time to recharge the precipitator can be more

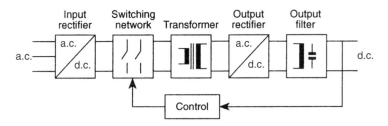

Figure 8.2   Block diagram of a typical SMPS system

controlled using smaller time step increments to speedily re-establish optimum precipitator performance.

8.2.3 Design considerations

The development of a switch mode power supply presents the designers with various challenges from the mains input topology through to the final rectification stage, plus the problems of controlling the device such that optimum precipitator performance can be achieved under the operational inlet conditions.

In designing a next generation switch mode precipitator power supply the first objective is to establish the exact requirements of the supply and ensure that the chosen topology will satisfy the following key areas.

(1) The operating frequency, voltage and current levels required by the system are obtainable using readily available tried and proven power semiconductor devices.

(2) Operating flexibility: the ability of the topology must be responsive to any required control variations in order to cater for changes in the precipitator inlet parameters.

    (i)   The voltage (or current) must be capable of being varied, either automatically or manually, following programmable ramps and timings, which could vary from fractions of a millisecond to several seconds depending on the prevailing conditions within the precipitator.

    (ii)   When a flashover occurs within the precipitator, the power supply must be capable of detecting it and interrupting the power flow rapidly to minimise the amount of energy lost. The system must then be capable of recharging the precipitator to a specified level with programmable times and methods.

(3) High operating efficiency. High power factor operation and good quality sinusoidal input currents are a pre-requisite of any modern power electronic installation. Poorly designed topology could easily exceed the limits on current harmonics set by European Standards. Additionally, a poor quality input current waveform could adversely affect the power supply circuit as follows:

    (i)   the power available from the equipment could be significantly reduced,

    (ii)   filtering components could be severely stressed because of large peak pulse currents and

    (iii) power losses in the system are increased as a result of high current pulses.

It will therefore be necessary to ensure that the precipitator power supply operates at a near unity power factor with high quality sinusoidal input line currents.

(4) Finally, the topology should be simple and the chosen approach must be cost effective.

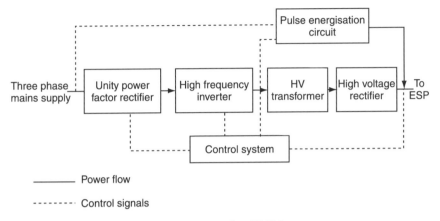

Figure 8.3   Generic component arrangement of an SMPS system

8.2.4 SMPS circuit configurations

Although the actual circuit design/configuration and components will depend on the supplier's preference to a certain extent, the basic equipment arrangement comprises the basic components as indicated in Figure 8.3.

(a) Rectification of the incoming three phase supply.
(b) The subsequent d.c. feeds an inverter, to generate a controllable square wave output.
(c) This provides the necessary primary supply to the HV transformer.
(d) The HV output is then rectified and feeds the precipitator.
(e) A control and monitoring system.
(f) Possible 'pulse energisation'[1] capability.

Although the figure indicates a typical arrangement, the various components can have different architectures depending on the supplier, and these can impact on the design and operation of the equipment as will be examined in the following section.

8.3 Input rectification stage

These are the requirements of the input stage of the power supply.

• This is basically an incoming mains rectification unit, which can provide a low ripple d.c. link voltage and should draw low distortion input line currents from the supply utility.
• The overall circuitry and components should be robust and reliable in use, and be produced at an economic cost.

---

[1] In effect, this is the application of a short duration wave (e.g. typical pulse width 50–100 μs) supplied by the inverter feeding the HV transformer.

- Ideally it should have a power factor approaching unity such as to minimise electromagnetic interference (EMI) problems and harmonic distortion of the supply.
- To minimise the size and weight of the high voltage high frequency transformer, it may be advantageous to have an input stage topology which is capable of boosting the d.c. link voltage in excess of the nominal 560 V d.c. as derived from a standard bridge arrangement. If this approach were adopted, an ideal target link voltage would be 800 V d.c., which, while remaining well within the limits of current semiconductor switch technology, would add cost to the system and introduce further power losses.
- The overall circuit efficiency must be high and, if possible, but not critical, should be capable of energy recovery to minimise power losses, particularly when 50–100 μs pulse widths are being used to mitigate high resistivity dust reverse ionisation difficulties.

From the above requirements of an electrostatic precipitator power supply, it was concluded that there were three possible circuit topologies for the input stage of an SMPS power supply:

(1) three phase six switch unity power factor (UPF) converter;
(2) three phase boost type unity power factor rectifier;
(3) three phase full wave rectifier, with a.c. side filtering.

8.3.1 Three phase six switch mode UPF converter

This circuitry is illustrated in Figure 8.4. This is a well proven approach and is essentially a four quadrant switch mode inverter. By pulse width mode (PWM) switching of the semiconductor devices, usually IGTBs, the converter is capable of drawing sinusoidal input currents at near unity power factor, which will reduce the EMI and distortion problems [2]. Figure 8.5 indicates a PsPice

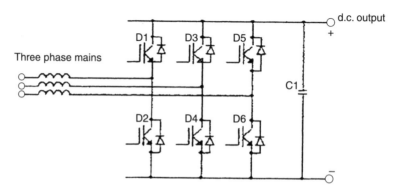

Figure 8.4   Three phase six switch mode UPF converter circuitry

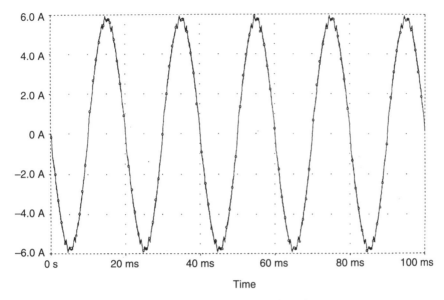

Figure 8.5   PsPice simulation of a three phase six switch mode UPF converter circuit

waveform simulation of the topology, indicating minimum distortion of the line current. The power flow through this topology is a direct function of switching control, therefore reversal of power flow is possible.

The cost of this topology is high because of the larger number of semi-conductor switching devices (6) and the heat sink requirements of these switches. The number of switching devices in this topology also means that a complex switching and control system will be required further inflating the cost.

8.3.2 Three phase boost type UPF rectifier

The circuitry for the three phase boost type of unity power factor (UPF) rectifier is illustrated in Figure 8.6. Again it is fairly widely used and hence is a reliable option for an input device. By control of the switching frequency and duty ratio of the semiconductor switch, the rectifier can be made to draw nearly sinusoidal input current from the utility with a power factor approaching unity, hence EMI and distortion is minimised [3]. Figure 8.7 shows a PsPice waveform simulation of the line current and voltage indicating minimal distortion of these waveforms. As the name suggests this topology is also capable of boosting the output d.c. voltage to a pre-determined level.

This topology has the advantage that it achieves all that the switched mode converter can offer (except for bi-directional energy flow) with just one switching device and one fast recovery power diode instead of six used for the previous system.

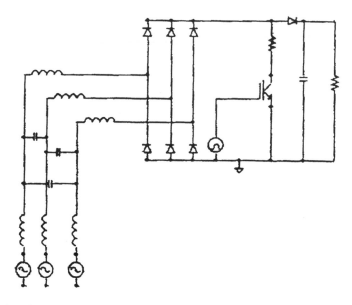

Figure 8.6   Three phase boost type UPF rectifier topology

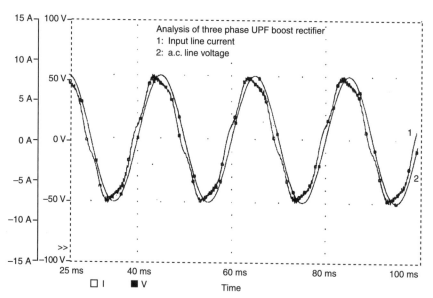

Figure 8.7   PsPice simulation of a three phase boost type UPF rectifier circuit

8.3.3 Three phase full wave rectifier with a.c. side filtering

The basic circuitry for this topology is illustrated in Figure 8.8. Although the approach is widely used in some current SMPS designs being relatively low cost, it can create harmonic distortion within the power distribution network and give

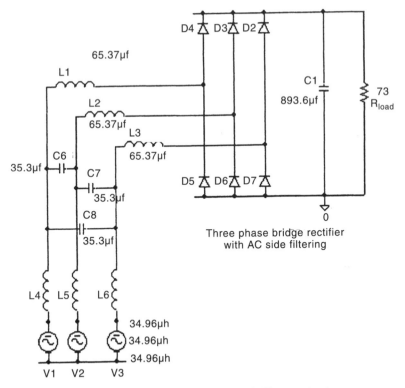

Figure 8.8 Three phase full wave rectifier with a.c. side filtering circuit

rise to higher peak currents. Figure 8.9 presents a PsPice simulation indicating the large distortion of the line current with this approach. Unlike the initial topology this system is not capable of bi-directional energy flow; however, this is not considered a serious drawback to its usage as the system is very robust and reliable, because only diodes and not switching semiconductors are employed in the circuitry.

8.3.4 Comments on the various input stage topologies

Based on the criteria upon which these particular topologies have been considered and analysed, the 'three phase boost type UPF rectifier', although the most expensive, has several potential operational advantages over the 'three phase six switch mode UPF inverter' and the 'three phase full wave rectifier with a.c. side filtering' approaches, as follows.

(a) It is capable of unity power factor (UPF) operation and boosting the voltage with the minimum number of switching devices and hence should be more reliable at a lower cost than the three phase six switch approach.

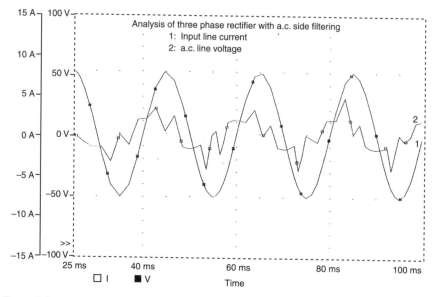

Figure 8.9   PsPice waveform simulation of a three phase full wave rectifier with a.c. side filtering circuit

(b)   The ability to boost the d.c. link output voltage can be advantageous because it reduces the turns ratio requirement of the high voltage transformer and hence reduces harmonic distortion, but at an additional cost over the other topologies.

(c)   Although bi-directional energy flow is only possible with the three phase boost topology, this is not envisaged to be a major problem with other systems, since the pulse energy could be fed back to the d.c. link rather than to the utility.

8.4   High frequency inverter stage

This stage must be capable of converting the derived d.c. line supply to a.c. in order to feed the high frequency, high voltage transformer such as to achieve a high performance operating level. Consequently the device must satisfy the following criteria:

- the maximum operating voltage and frequency must be achieved using commercially available devices at high reliability and acceptable cost, and
- the system must be flexible and easy to control.

The operating frequency in an inverter feeding a high voltage transformer is controlled not only by the semiconductor switching losses, but also by the reactance of the transformer itself. With the high voltage element requiring reasonably thick insulation between the windings, this can produce a high leakage inductance and the large number of turns in the secondary can lead to a high distributed value of capacitance between the coil layers. This capacitance is reflected into the primary as a parasitic capacitance, which can limit the operational frequency of the inverter.

In reviewing the requirements listed above, there are three possible topologies that can be considered:

(1) PWM controlled 'H' bridge inverter,
(2) resonant converter, and
(3) matrix converter.

8.4.1 PWM controlled 'H' bridge inverter

A typical PWM controlled 'H' bridge inverter circuit is shown in Figure 8.10. which is an example of the current technology being used by some switched mode power supply manufacturers, and it is a well proven and reliable method of achieving the required switching waveform for the HVHF transformer. The inverter operates in square wave switching mode, where the average output voltage is controlled by pulse width modulation (PWM) and where the instantaneous peak voltage is controlled only by the d.c. link voltage. There is, however, a possible problem with this topology in that if the parasitic elements within the HVHF transformer are not taken into account these can limit the self-resonant frequency of the transformer and subsequently the maximum working frequency in PWM mode. Furthermore, the parasitic elements result in high stresses and losses within the switching devices themselves. In practice, to overcome these losses, this topology needs to be operated at a reduced switching frequency, which requires a significant increase in the size and mass of the HVHF transformer [4].

8.4.2 Resonant converter

During recent years a large variety of resonant switched mode converters have been introduced, which has led to a situation where it is difficult to classify the various resonant converter based topologies. Figure 8.11 presents a general

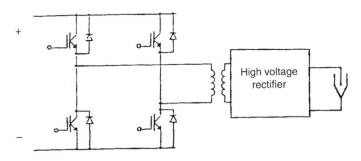

Figure 8.10   Typical PWM controlled 'H' bridge inverter circuitry

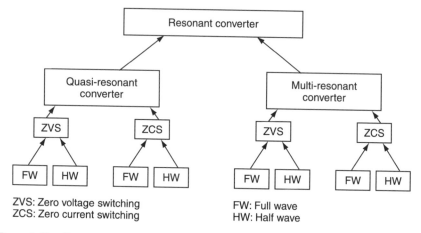

Figure 8.11   Resonant converter classification

classification scheme for some typical resonant converters. Although subtle variations in operation and performance exist for each particular topology, the general theory of resonant converters holds true for all. The resonant frequency of the inverter may well be tens of kilohertz and the output voltage (or current) is controlled by varying the switching frequency, either above or below the resonant point, dependent on the topology selected.

Some conventional d.c. link inverters use 'hard' or 'stressed' switching, i.e. the semiconductor switching devices turn on and off at the full d.c. link voltage. This results in high switching losses in the semiconductor, which can limit the inverters' operating frequency, voltage and current levels. With some supplier preferred designs of circuitry, the resonant circuitry can be found located either at (a) the input of the inverter, (b) the output of the inverter or (c) across the switching devices themselves. The purpose of the resonant circuitry is to convert the fixed link d.c. voltage (or current) to a pulsating form.

An opportunity arises with this topology where it is possible to turn on or off the switching devices during the zero voltage or zero current crossing point. Zero voltage or zero current switching is often referred to as 'soft' switching, because turn on or turn off losses disappear, resulting in a high inverter efficiency [5].

The resonant converter using soft switching offers the following advantages.

- The device switching losses can partly or fully disappear by using soft switching, thereby resulting in a high inverter efficiency.
- Because the switching losses can be reduced the heat sink requirements are minimised. This is of particular importance at high power levels where the heat sink costs can be significant.
- The converter can operate with minimal snubber circuitry.
- Semiconductor device reliability is improved because the stress due to voltage and currents excursions are reduced.
- EMI (electromagnetic interference) problems are less severe because the res-

onant voltage pulses have lower dV/dts compared to those of a standard PWM inverter.

• All the preceding factors make the inverter size small and relatively inexpensive.

• The reduced size and smaller heat sink requirements open up the possibilities of working at higher design power levels.

The advantage of the resonant converter compared with the standard PWM inverter is that the parasitic elements can actually be used to form the majority (or part) of the resonant circuit, thereby freeing the converter of any serious operating limitation.

8.4.3 Matrix converter

Figure 8.12 illustrates a matrix converter, which is an alternative to the standard 'H' bridge type inverter, and in this particular topology it is shown directly powering the HVHF transformer. The matrix converter basically provides a direct a.c. to a.c. conversion with an inherent unity power factor, but without an energy storage capability or d.c. link. Because the entire operation of a matrix converter is dependent on the switching algorithm being used, the input currents can be controlled to be sinusoidal and in phase with the mains supply voltages [6, 7].

Although the matrix converter is widely used for low frequency motor drive applications, in order to operate at high switching frequencies (>20 kHz), the output waveform will essentially become a high frequency square wave. To control and modulate this waveform while at the same time protecting the switching

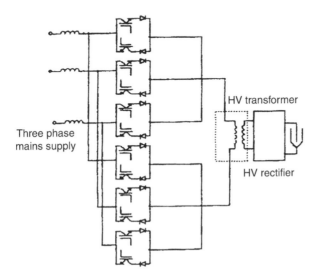

Figure 8.12   Matrix converter topology

devices against overvoltages requires a complex and high speed switching algorithm.

8.4.4 Comments on the various inverter topologies

Although there are potentially three types of inverter topology that can be applicable to the design of switched mode power supplies, the matrix type converter is at present unproven, whereas both the 'H' bridge and resonant converters have been successfully used by different equipment suppliers. Both these approaches have been used satisfactorily dependent on the basic design topology of the associated circuitry. Although the 'H' bridge has the simplest overall topology, it has an operational frequency limitation resulting from the HVHF transformer parasitic elements.

8.5 High voltage, high frequency transformer

The design of a high voltage, high frequency (HVHF) transformer differs widely from the standard transformer design methodologies used for mains voltage applications. These differences need to be addressed if the design of the HVHF transformer is to be viable [8, 9, 10]; these are:

- parasitic capacitance,
- magnetic leakage flux,
- insulation and electrical stress management,
- core losses and heat dissipation, and
- corona effects.

8.5.1 Parasitic capacitance

High voltage transformers used for precipitation duties usually have a large turns ratio, typically up to 200:1 for current applications, and, as a result, any small stray capacitance values existing between the secondary winding layers can produce a significant overall effect. When referred to the primary, this reflected capacitance, which is proportional to the turns ratio squared, takes the form of a significant parasitic element. As this capacitance can affect the operating performance and efficiency of many switching inverter topologies, it necessary to take steps to minimise this potential problem in the design. One method is to split the secondary winding into a number of interconnected windings individually separated by an insulating layer, which has the effect of serially connecting the individual capacitances and hence reduces the overall impact.

8.5.2 Magnetic leakage flux

The magnitude of leakage flux within the HVHF transformer will be determined by several factors.

- A high coupling factor between the primary and secondary winding will minimise the leakage flux.
- The core shape and aspect (height to width) ratio will affect the amount of leakage flux.
- Any sharp corners through which the magnetic flux has to pass will result in flux leakage.
- The magnitude of the flux within the core affects the total amount of leakage flux.

Core losses can be reduced by having a suitable low loss ferrite material in the basic design. This material must be capable of operating at a high operational frequency and avoid magnetic saturation, by using for example, a high saturation manganese–zinc ferritic material [11]. The power loss density of the material is <0.1 W cm^{-3}, at an operating frequency of ~25 kHz and a high flux level, plus it also exhibits a reducing loss with increasing temperature, particularly towards the Curie temperature. The physical size and structure of the ferrite must also be such that it is capable of dissipating any heat generated in the core.

8.5.3 Insulation and electric stress management within the transformer

The efficiency of a transformer integrated into a power supply capable of delivering up to 100 kVA is of prime concern, because even at an efficiency of 99 per cent, the losses will result in significant heat being developed within the transformer and the dissipation of this heat to the surroundings is clearly an important design issue.

One of the incentives for adopting high frequency switched mode technology is to minimise the size and weight of the transformer and a factor influencing its size is the maximum electric stress that can be tolerated by the insulating media used in the design. Oil insulated transformers are heavy and prone to leaking and the oil needs to be regularly cleaned – see Chapter 5. Otherwise high stress-points can build up as a result of any impurities within the oil, which can lead to electrical breakdown; however, in spite of these difficulties there are a number of operational SMPS designs using oil insulation.

Inert gases generally have a dielectric strength some two to three times higher than that of air; hence, a method of reducing the required insulation gap between primary and secondary windings is to encapsulate the transformer in a sealed unit containing an inert gas at a pressure slightly above atmospheric. An advantage of this approach is that these gases can have thermal transfer coefficients up to three times greater than air, which greatly assists in heat dissipation from the core and windings.

Solid insulation may be used, but it can give rise to difficulties during transformer development (if breakdown occurs during proof testing of the transformer it may be necessary to start again from square one) and solid insulation can restrict heat transfer. Some solid insulation is, however, required for mechanical support of the secondary windings in the final design.

8.5.4 Corona effects

Any corona discharges that might develop can seriously affect the operation and life expectancy of any high voltage transformer unless prevention methods are integrated into the design. Any sharp corners or rough surfaces within the magnetic structure of the transformer can lead to a high electric field stress and potential corona discharge. This corona will create highly reactive molecules, which can, in time, degrade the electrical insulation leading to complete electrical breakdown. It is vital, therefore, to design the magnetic structure such that all surfaces have a profile/curvature that minimises the potential for corona production.

Although the design of a high voltage, high frequency transformer is very different to that used for normal mains frequency operation, adopting the foregoing basic approach to the design, a HVHF transformer will prove reliable and meet the stated requirements.

8.6 Transformer output rectification

Because of the significance of adding capacitive components to the circuitry and its impact on the potential operational frequency of the system, the avalanche type of silicon diode, rather than capacitance/resistive voltage sharing, is commonly used for rectifying the transformer output. Although the capacitance/resistive sharing diode network can be used, the added capacitive effect on operational frequency must be taken into account in the design. In either case the effect/possibility of high ringing voltages arising as a result of flashover within the precipitator fields, similar to those experienced in the case of conventional mains rectifier equipment, must be considered in the design. Whichever approach is adopted in the final SMPS unit, the diodes must be designed for high frequency switching.

8.7 Short duration pulse operation

In the case of electricity generation plants firing coals producing high resistivity ash conditions giving rise to reverse ionisation, there has been a need to modify the conventional mains frequency supply to a pulse energisation technique to optimise and enhance the collection efficiency of the precipitator. This is generally achieved by superimposing a short negative pulse of high amplitude but variable frequency on a reduced negative d.c. base level, as discussed in Chapter 7.

There are several SMPS circuit topologies designed specifically for this task based upon charging and discharging the precipitator resonantly, the period of which is controllable. In the ideal case, the resultant pulse voltage is usually a perfect half wave sinusoid and all the energy is completely recovered at the end of the pulse. In practice, however, circuit performance will be limited by the switching and transformer losses; therefore, a topology that recovers pulse energy and feeds it back to the mains supply or d.c. link may prove to be a better alternative to conventional circuit arrangements.

In an alternate design, the converter switching is arranged to feed short duration 50–100 µs pulses into the HVHF transformer, rather than rely on resonance with the precipitator capacitance. This will produce a similar high pulse voltage to a conventional arrangement and can be followed by a lower pulse height to provide the base voltage requirement.

8.8 Advantages of the SMPS approach to precipitator energisation

In general, the real and perceived advantages of the SMPS approach to precipitator energisation can be summarised as follows.

(a) The unit will be much smaller and lighter than a conventional mains rectification unit.

(b) Being three phase driven, the unit requires lower current rated cables and switch gear, which has a considerable impact on prime and installation costs.

(c) The three phase connection enables a power factor approaching unity to be achieved and hence minimise mains harmonic distortion.

(d) Dependent on the cooling approach used, the unit can be designed to be 'mineral oil free', which reduces fire risks and associated installation costs.

(e) With the faster switching capabilities, e.g. 0.02 ms compared with 8.3–10 ms, electrical conditions on the precipitation field can be better optimised.

(f) Not only is the corona current input significantly increased, but the decrease in ripple voltage raises the mean operating voltage, both leading to performance improvements.

(g) The improved switching, where the line voltage is PWM controlled, reduces the dependence of the equipment on the line supply conditions and hence both EMI and harmonics are minimised.

(h) The switching allows for a wide variety of waveforms to be derived from a single SMPS unit, e.g. from pure d.c., through full intermittent energisation, to short (e.g. 50–100 µs) pulse operation. This means the full range of electrical operating conditions, arising with dusts having different electrical resistivities, can readily be met, by a simple change in switching procedure.

8.9 Review of the various topologies leading to a prototype SMPS development under a UK EPSRC grant

The following summarises the investigations carried out reviewing the design criteria for a switched mode power supply and the development of a prototype form of equipment under the auspices of an EPSRC Grant [12]. This development was undertaken using the assets and expertise of a team comprising a university, a commercial equipment manufacturer and an end user.

Although various topologies were examined, it was considered that having a boost type inlet stage and raising the line voltage to some 800 V d.c., from the normal 560 V d.c. from a conventional three phase bridge circuit, is advantageous in that, in addition to reducing the number of secondary turns required on the high voltage transformer and therefore assisting in reducing parasitic capacitance effects, it also produces almost a unity power factor with respect to the incoming supply.

As regards the inverter stage, possibly the simplest and cheapest is the 'H' Bridge approach, recognising, however, that this has a disadvantage in that the overall impedance of the circuitry can restrict the range of operating frequency.

With regard to the HFHV transformer, it was decided to limit the secondary turns ratio in order to reduce the parasitic capacitance component, thereby enabling the 'H' bridge to function close to its design frequency. To achieve the required voltage necessary to meet a full scale installation, two or more rectifier/transformer units would be connected in series, within the HV tank.

To produce a 'dry package', such as to eliminate oil for cooling and insulation, the unit used sulphur hexafluoride, under 2 bar pressure, for electrical insulation and heat transfer purposes. Although it was appreciated that the final unit may contain another inert gas, e.g. nitrogen, it was decided to investigate sulphur hexafluoride, because it has the highest electrical and thermal heat transfer properties of the inert gases.

This prototype unit, as illustrated in Figure 8.13, was field trial tested on a large power plant precipitator and its operation compared with that of the installed conventional 50 Hz rectifier unit. The results of this trial, although relatively short, confirmed the design as being satisfactory, in that both the mean operating voltage and currents were increased, which would have resulted in an enhanced precipitator performance, had all installed rectifier equipment been changed to SMPS units.

8.10 Operational field experience with SMPS precipitator energisation

Although the foregoing describes and reviews the various topologies that may be used for SMPS designs, the following section examines some of the commercially available SMPS units that have been built and operated in the field.

Figure 8.13   Photograph of the EPSRC prototype SMPS equipment (courtesy Leicester University)

Over the past decade a number of international companies have been developing high voltage, high frequency equipment and gaining field experience on a wide variety of processes, initially as potential retrofit applications but, more recently, with the benefits of SMPS operation, for new plants. The following summarises some of these investigations and their impact on precipitator performance. It will be shown that, as a result of SMPS energisation, the increases in mean operating voltage and corona current flow significantly improves performance over conventional forms of energisation. The increase in corona current is particularly beneficial when handling low resistivity particulates. In the case of high resistivity dusts, the system can operate under both intermittent and short 50–100 μs pulse mode forms of energisation with some-what greater benefits over mains units as regards performance enhancement.

8.10.1 SMPS system of Supplier No. 1

The approach adopted for the system topology is based on a conventional three phase bridge rectifier system, the output of which is rectified and carefully fil-tered by a smoothing circuit that includes a large buffer capacitor. The resultant output then feeds a series resonant converter connected to the transformer, the input being regulated by varying the switching frequency above the resonant frequency. The output of the HVHF transformer is then rectified by a monopha-sic full wave diode bridge to supply the precipitator field. The heart of the system is the d.c./a.c. converter, implemented as a full bridge series resonant converter, comprising four switches, a series capacitor and a series inductance, each switch being implemented with a semiconductor IGTB device. The basic architecture of this switched integrated rectifier system is illustrated in Figure 8.14.

8.10.1.1 Initial field trials

The initial trials of this equipment were carried out on a wood chip fired boiler unit producing a dust having a lowish resistivity of 10^{10} Ωcm [13]. The precipita-tor under test comprised two fields and the SMPS equipment, rated at 80 kV, 250 mA running at 50 kHz, was installed in parallel with one of the existing TRs through a changeover switch. This arrangement allowed comparison to be made between normal conventional energisation and a high voltage, high frequency energisation system.

The results from these initial field trials showed that with the faster charging rate of the SMPS system, 0.02 ms compared to 10 ms with conventional energi-sation, the ripple voltage was significantly reduced. This resulted, not only in the corona current increasing from 110 mA to 220 mA, but also the mean operating voltage rose by some 10 kV. These changes in active power input producing a significant enhancement in precipitator performance.

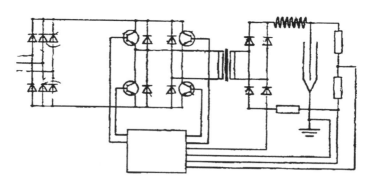

Figure 8.14   Basic circuitry for SMPS system No. 1

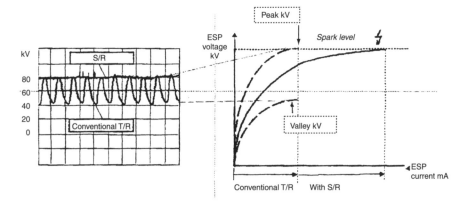

Figure 8.15   Comparison of voltage waveforms from conventional and SMPS energisation systems

Figure 8.15 shows waveforms comparing the voltages obtained with SMPS operation and conventional energisation [13]. Although the sparking/flashover voltages are similar, the large ripple voltage clearly illustrated with conventional energisation results in a significant reduction in the mean operating voltage, whereas that from SMPS energisation is almost a pure d.c.

Figure 8.16 illustrates that with the improved control available with the SMPS system, the time 'lost' in recovering optimum electrical conditions following a flashover is reduced, as is the 'quench' time, leading to an overall enhanced performance being achievable with SMPS energisation.

8.10.1.2   Subsequent installations and performance enhancements

As a result of the field trials and laboratory investigations, Supplier No. 1 has developed a range of high voltage high frequency supplies, with ratings up to

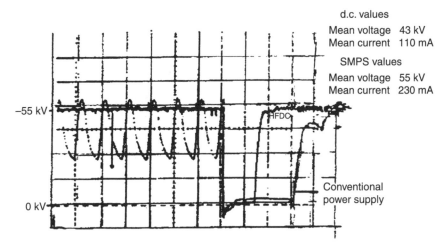

Figure 8.16   Reduction in electrical recovery times following a flashover

100 kW. These units have been installed on mist, wet and dry type precipitators in various parts of the world, the dry precipitators handling both medium and high resistivity dust applications.

In the case of low resistivity tar mist, the replacement of the conventional equipment by SMPS energisation reduced the emission from 40 mg Nm^{-3} to 5 mg Nm^{-3}. In general, the application of SMPS has been found to result in at least a 40 per cent reduction in the emissions obtained using conventional energisation. Although SMPS does not alleviate space charge effects, the application of SMPS to both the inlet and second fields of a precipitator, with the higher corona current inputs, serves to reduce the impact of space charge to a certain extent, thereby enhancing overall performance.

Table 8.1 presents a summary of the various applications and data obtained following retrofitting or supplying new equipment with SMPS energisation by Supplier No. 1 [14].

The switched integrated rectifier unit SIR, developed by Supplier No. 1, is a modular, self-contained system, comprising the incoming switch gear, high voltage power pack and control system, including motor control facilities for rapping, heating, etc., all mounted in one box having side dimensions of around 800 mm. The weight of the complete system is some 200 kg, which can be installed directly onto an ESP HT bus duct, the unit then only requiring a three phase cable connection. A photograph of the SMPS equipment is shown on the right in Figure 8.17, in comparison with a conventional mains frequency rectifier plus control panel of the same output capacity.

8.10.2 SMPS system of Supplier No. 2

Although not having as much field experience, the high frequency switching system developed by Supplier No. 2 uses a series resonant circuit type converter; otherwise the circuit is fairly standard. The power from a three-phase mains

Table 8.1 Typical improvements attributable to SIR operation on various ESP applications

Application	Location	No. of SMPS	New or retrofit	Emission reduction
15 Boiler plants	World-wide	70	Retrofit	Various levels
Boiler plant	USA	32	Retrofit	Met emission
Boiler plant	USA	12	Retrofit	60%
26 Soda recovery	Scandinavia	74	Retrofit	40–60%
19 Wet ESPs	World-wide	44	Both	40–85%
15 Cement plants	Europe	44	Retrofit	Up to 75%
10 Waste to energy	Europe/Japan	28	Retrofit	20–45%
49 Biomass units	Scandinavia	75	Both	10–50%
12 Glass industry	USA/Europe	51	Both	Up to 60%
15 Miscellaneous	World-wide	40	Retrofit	Various levels

Figure 8.17   Photograph of SMPS unit in comparison with a conventional TR and control cabinet of 70 kV, 800 mA output (courtesy Alstom Power)

supply is rectified, then filtered through a smoothing circuit which includes a buffer capacitor. The outgoing pure d.c. voltage then feeds a series resonant converter, operating at around half the resonant frequency, connected to the HFHV transformer; the secondary voltage, after rectification by a monophasic diode bridge, is applied directly to the discharge electrodes of the ESP. The real heart of the system is the d.c./a.c. converter, implemented as a full bridge series resonant converter, comprising four switches (S1–S4), a series capacitance C_s and a series inductance L_s. Each switch is implemented with a semiconductor device (IGBT) and an anti-parallel diode. The architecture of the system is indicated in Figure 8.18 [15].

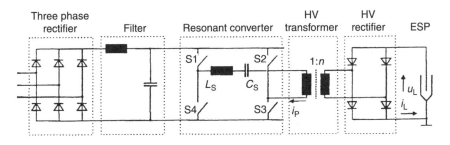

Figure 8.18   SMPS circuit architecture for Supplier No. 2

The values of L_s and C_s, and the turns ratio (n), are chosen in such a way that the amplitude of the current pulses is practically independent of the capacitance of the ESP; that is, this approach operates as a constant current generator. This is an advantage compared with the traditional equipment, because operation with the SMPS system eliminates the current surges due to sparks and arcs arising inside the ESP. In case of an ESP short circuit, the primary and secondary currents retain their normal amplitudes and the line current supplied by the mains falls to a very low value, corresponding only to the power losses within the SMPS system. With traditional equipment, the line current is limited by a linear inductor and total circuit impedance, to between two and three times the rated primary current. In other words, with SMPS, the mains supply is very insensitive and basically isolated in respect to the incidence of sparks, possible arcing and short circuiting within the ESP.

8.10.2.1 *Initial field trials*

The prototype SMPS equipment used during the following test is rated at a mean current of 600 mA and a maximum voltage of 100 kV, which is similar to the original mains equipment used on this particular application, hence a comparison test performed with the two power supplies is acceptable, as both conventional and SMPS units are rated to the same output mean current. A photograph of the SMART (switched mode advanced rectifier technology) unit is shown in Figure 8.19.

The ESP used for the prototype SMPS test dedusts the waste gases from a dry process cement installation. This comprises a short rotary kiln followed by a four stage cyclone preheater system rated for a clinker production of 1500 t each day, together with a raw meal mill. When the raw mill is stopped, the kiln gases are cooled in a conditioning tower operated in parallel with the raw mill. The raw mill is stopped for about 7 h every day to reduce energy consumption, and during this period it is possible to promote back-corona by increasing the gas temperature in the ESP by reducing the water injected into the conditioning tower. This enabled comparison to be made between medium and high resistivity dust conditions for both forms of energisation [16].

The ESP consists of two flows (A and B) each with two series fields, the outlet of each flow being connected to its own stack enabling direct emission comparisons to be achieved. The main test was performed on flow A, the SMPS prototype being connected to the outlet field A2. Although two series of tests were carried out, one at full kiln output and another at around 80 per cent loading, only the full load test data will be reported.

8.10.2.1.1 *Low resistivity dust conditions*   With the kiln operating at a full production rate of 90 tph, tests were carried out under precipitator inlet conditions with water injection at 3 m^3 h^{-1}, giving a low dust resistivity with a gas temperature to the ESP inlet of around 130 °C.

Figure 8.19   Photograph of the SMART SMPS unit (courtesy FLS Miljö als)

Power supply	Mode	Jmean mA	Umean kV	Upeak kV	Umin kV	Emission mg Nm⁻³	Power kW	Power kVA	PF
T/R	d.c.	550	47	61	37	81	34	44	0.76
SMPS	IE*	490	46	71	37	73 (94)	33	36	0.91

*t_{on} 1.7 ms; t_{off} 5 ms

The emission figure shown in parenthesis corresponds to the higher emission obtained with SMPS operating under pure d.c. energisation.

Under these low resistivity dust conditions, conventional d.c. energisation was found to be better than intermittent energisation, however, with SMPS operation, intermittent energisation was found to give lower emissions than for pure d.c. energisation.

8.10.2.1.2 Moderately high dust resistivities Similar comparative tests were carried out under moderate back-corona conditions, with the operating conditions derived from an ESP inlet gas temperature of around 180 °C.

Power supply	Mode	Imean mA	Umean kV	Upeak kV	Umin kV	Emission mg Nm⁻³	Power kW	Power kVA	PF
T/R	IE*	275	36	63	25	138	16	33	0.48
SMPS	IE**	420	39	66	26	118	24	26	0.91

*Nec = 3 and **t_{on} 1.7 ms; t_{off} 6 ms

Under these moderate to high resistivity conditions, IE with both modes of energisation were found to produce the lowest emission or highest performance level. The reason for the lower emission under SMPS operation is because of the higher ESP corona power input. An important finding from the test is the significantly higher power factor achieved with SMPS operation.

This result from this initial prototype test indicates that SMPS energisation can improve the ESP collection efficiency for both low and moderate dust resistivities, the improvement being greatest under conditions of back-corona. The best SMPS operational mode was intermittent energisation, which was always superior to pure d.c. energisation, as a consequence of the higher voltage levels obtained. Figure 8.20 illustrates the waveforms obtained with various energisation modes [16].

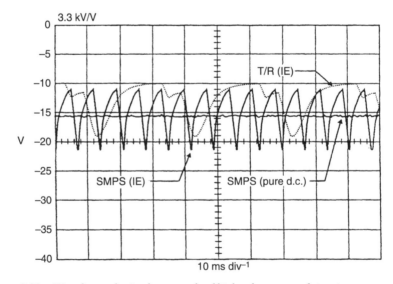

Figure 8.20 Waveforms obtained as a result of high voltage rate of rise times

8.10.2.2 Subsequent proving trials

A second test series has been carried out by Supplier No. 2 on a Leca Kiln having a design volumetric pellet production rate of 55 m³ h⁻¹ [17]. The ESP, which was in a good mechanical condition, consists of only one flow with two series fields having a total collecting area of 1080 m². Each field is energised by a 200 mA/90 kV conventional T/R set controlled by a PIACS d.c. control unit. Because of very favourable market conditions, the plant can sell all its production; consequently the kiln output has been increased significantly, which has resulted in a 75 per cent increase in the design ESP gas flow and produced an unacceptable increase in the particulate emission of up to some 140 mg Nm⁻³. The emission situation has been compounded by the high dust resistivity of up to 5×10^{12} Ωcm at 190 °C. The reason for this high resistivity is the presence of organic substances in the layer on the collecting plates originating from the materials used in the clay/oil blend, fed to the kiln.

For these tests, two SMPS units were installed on the precipitator roof and connected to the respective field through a switch box, which allowed for changeover and comparisons to be made between the existing T/R and SMPS forms of energisation. The arrangement is illustrated in Figure 8.21. The figure also indicates how the SMPS system microprocessor based control units have a serial communication bus used for downloading the data to a supervisory PC installed in the control room of the plant. This supervisory unit has a built-in modem and suitable software allowing remote operation of the installation via a telephone line, which was used for remote long term assessment of SMPS

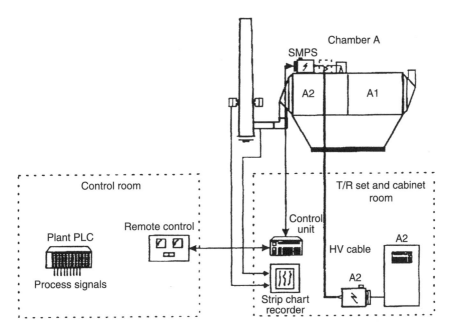

Figure 8.21   Arrangement of equipment for SMPS assessment testing

operation. The ESP dust emission is recorded by a calibrated Sick OMD 41 Opacity Monitor mounted on the outlet ducting after the ESP. This opacity signal is also fed to the supervisory monitoring system, as is the ESP flue gas temperature signal.

The preliminary short term comparison tests produced a minor improvement in favour of SMPS energisation, the dust emission being typically 8–9 per cent lower.

	Upeak kV	Umean kV	Imin mA	Imean mA	fspark spk min^{-1}	$t_{on\ max}$ ms	t_{period} ms
SMPS$_1$	104	25	17	32	4	3	60
SMPS$_2$	102	14	11	2	1	2	500

	Upeak kV	Umean kV	Imin mV	Imean mA	fspark spk min^{-1}	t_{limit} ms	Nec (half-cycle)
T/R$_1$ IE	74	26	17	41	0	41	5
T/R$_2$ IE	73	13	10	14	0	14	19

Nec is the degree of intermittence or charge ratio.

The SMPS and conventional T/R sets were operated under optimised electrical operation producing the voltage and current levels shown in the above table. Owing to the high dust resistivity, IE energisation was used throughout the programme; the degree of intermittence, however, especially in the outlet field, is significantly higher than one would normally anticipate for this type of dust, this rate leading to extremely low current levels. The peak voltage obtained with the SMPS system is, however, considerably higher and was obtained with a rate of rise of 150 kV ms^{-1}. Figure 8.22 presents an oscillogram of the voltage waveform of the inlet field. The power consumption with both energisation modes is similar and quite low, approximately 3 kW for the inlet and 1 kW for the outlet fields.

The most interesting result of this particular investigation was the long term comparison between the two energisation modes using the modem/telephone link, starting with an optimum initial reference emission under SMPS energisation of 89 mg Nm^{-3} wet, which, immediately following the change over to T/R energisation, increased slightly and continued to rise slowly during the next 6 h, reaching a maximum value of 147 mg Nm^{-3} wet. At this point energisation was switched back to SMPS operation, and immediately after the changeover the emission remained almost constant, but slowly started to fall and after 6 h had dropped to 104 mg Nm^{-3}, wet. Overnight the emission decreased further to

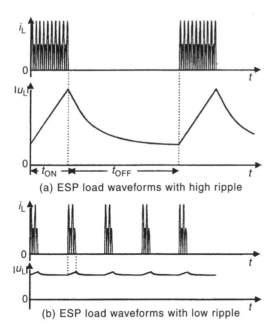

Figure 8.22   Waveforms obtained with normal and SMPS high rise rate operation

90 mg Nm^{-3}, wet, i.e. the starting value. This test is illustrated by the data shown in Figures 8.23(a) and 8.23(b), respectively.

Based on these results, it was decided to continue with long term comparisons performed remotely from head office using the telephone/modem link. These results, in terms of comparative emissions, are presented in the following table, covering an operating period of some 9 months.

Series No.	No. 1 July/Aug 99		No. 2 Nov/Dec 99		No. 3 Feb/Mar 00	
	T/R	SMPS	T/R	SMPS	T/R	SMPS
Power supply						
Dust emission	143–150	106	120–124	94	124–128	90

All values are given in mg Nm^{-3}, wet.

The findings from these investigations indicate that the use of SMPS, as a high voltage power supply for ESPs, is a very attractive technical solution because of its inherent advantages. These can be summarised as having less weight and volume than a conventional T/R because of the high switching frequency used, and as a result of its three phase nature and high power factor (above 0.9) there is an improved load situation on the public utilities. The system also proves a more flexible power supply because of its independence of the line frequency

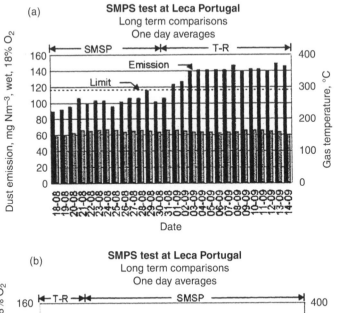

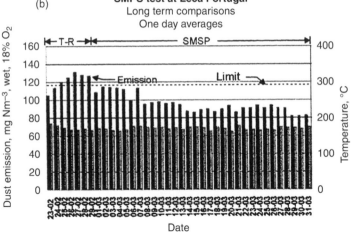

Figure 8.23   Long term opacity monitoring during different energisation modes

and can readily provide voltage waveforms from a pure d.c. to a variety of intermittence formats. These waveforms enable the installation to cope with any particular dust or operating condition such that a higher precipitator performance can be obtained.

8.10.3 SMPS system of Supplier No. 3

This supplier has developed an SMPS system so as to enable it to be split into a low voltage driver unit, containing the incoming switchgear, d.c./a.c. inverter and all control equipment, and the second unit, a self-contained high voltage system. This arrangement for a 70 kV 500 mA rated unit eases transportation, the respective weights being only 50 and 25 kg, and also facilitates mounting of the HT unit directly onto the precipitator bus duct connection. The block diagrams for both units is presented in Figure 8.24, while photographs of the equipment are shown in Figures 8.25 and 8.26 [18].

The main difference in the basic design of this approach, other than being a split unit, is the use of 'hard switching' of the inverter, as opposed to the more normal 'soft switching' approach. The reasons put forward are related not only to the lower number of circuit components through which switching losses arise, but, by substituting resistive elements by capacitive/inductive components whereever possible, this further reduces losses and hence minimises the cooling requirements of the system.

Hard switching is considered ideal for SMPS system operation, because unlike conventional control, where the thyristors operate in their linear region, in SMPS approaches, the power devices are either 'on' or 'off'. In the 'on' condition, the current is high with almost zero voltage drop, but reversing in the 'off' condition to almost zero current but high voltage drop. The extremes of zero current or zero voltage means that the power as the product of current × voltage approaches zero, so the power drop across the device is negligible.

To optimise these advantages, the circuit design includes a 'Buck type', low output impedance voltage converter [19], rather than a resonant type. The Buck converter, although possibly having slightly higher switching losses, offers an advantage in that the output voltage can be readily controlled by PWM switching without the need of complex feedback control circuitry. The PWM switching control minimises/limits the power supplied to the precipitator in the event of heavy flashover or severe arcing situations, thereby further protecting the components.

To minimise EMC noise problems arising from hard switching at high current, the main circuit components are connected through a built in 'heavy duty bus bar printed circuit board (PCB). The bus bar minimises the length of the connections and, being short, compact and fairly large, in terms of cross section, results in very little parasitic inductance which reduces potential EMC problems.

When working with intermittent energisation (IE), for dealing with high resistivity particulates, the rate of voltage rise is critical if optimum precipitator performance is to be obtained. Having a high dV/dt with a short time frame enables the peak precipitator voltage to be increased above the 'normal' breakdown level, without flashover occurring and without running into reverse or back-ionisation problems, which enhances the precipitator collection efficiency. The maximum value of dV/dt is limited by the maximum current capability of the supply. Because all precipitators have a significant capacitance (>50 nF) and as $I = C\,dV/dt$, in order to achieve a dV/dt of some 25 kV ms^{-1} requires a peak current of 1.25 A [18]. If higher values of dV/dt are required then the supply must be capable of delivering larger peak currents. With this equipment, this is

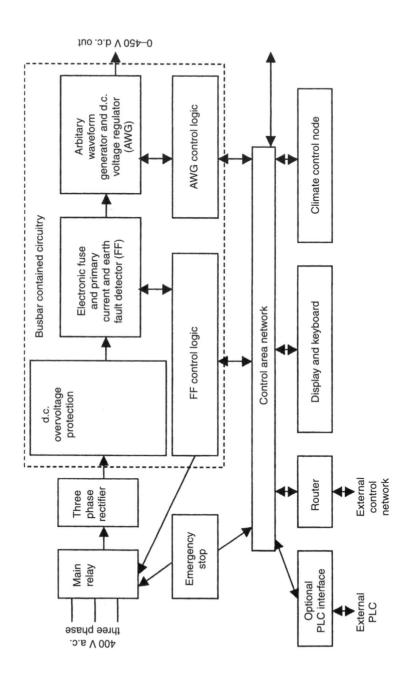

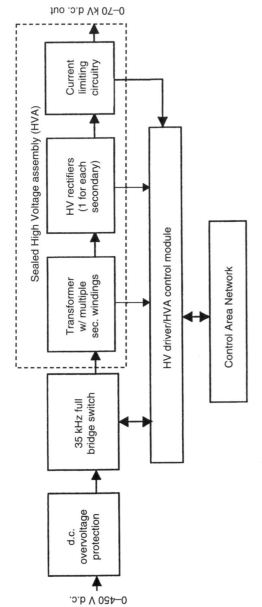

Figure 8.24   Block diagram of the driver and HT unit

Figure 8.25   Photograph of the control equipment (courtesy Applied Plasma Physics)

Figure 8.26   Photograph of the HT equipment (courtesy Applied Plasma Physics)

achieved by changing the function of the Buck converter from a constant voltage to a constant current source by simply changing the active control function of its internal circuit, thereby eliminating the need of a complex feedback loop. As the current is controlled purely as an internal parameter there is no influence on the control mechanism by load conditions and in consequence the active control function acts instantly [18]. Using this approach enables currents of around 2000 mA to be obtained from a supply rated at 500 mA, which enables dV/dts well in excess of 30 kV ms^{-1} to be readily achieved.

In summary, although the approach used in the design of this SMPS equipment is slightly different to that used by other equipment suppliers, it nevertheless appears to have be well thought out and is reported to perform satisfactorily with in-house precipitator applications. At present, full scale tests are being undertaken in the field on a large wet precipitator installation on a power station gas clean up system.

With its lighter waterproof construction it should become a strong candidate, initially for the retrofit market, but eventually for use on new installations. For new plant projects, the increased performance levels obtainable with this form of energisation should result in a smaller and lower cost precipitator being installed to meet a specific duty.

8.10.4 SMPS system of Supplier No. 4

The design of the equipment from Supplier No. 4 is a fairly conventional three phase rectifier system producing a d.c. line voltage of 650 V from a 480 V supply. This d.c. voltage is then passed through a resonant converter containing integrated gate bipolar transistors (IGTBs) to produce a 25 kHz, high frequency a.c., which is then transformed and rectified to feed the precipitator. A block diagram of the system is illustrated in Figure 8.27, and a photograph of the equipment is shown in Figure 8.28 [20].

To minimise stray capacitance and other transients, the circuit requires that the a.c./d.c. and d.c./a.c. modules are located close to the step up transformer, hence a fully integrated design was developed. With the faster switching capabilities of the IGTBs, the circuit can be made to operate some 250 times faster than a conventional system, which can only be switched at the zero crossing point of the incoming voltage. This fast switching feature enables benefits of

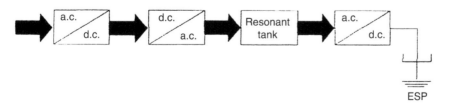

Figure 8.27   Block diagram of the SMPS system

Figure 8.28   Photograph of the equipment (courtesy NWL Transformers)

intermittent or short duration pulse energisation modes to be realised with a suitable control algorithm.

The SMPS system is based on a self-contained unit, which can be mounted directly on the roof of the precipitator without the need of cabling or bus ducting. The overall size of a 70 kW unit, including a manual earthing switch, is roughly a 1 m cube, weighing 340 kg and containing some 100 l of fluid.

In operation, the high frequency derived d.c. has a ripple voltage of some 3–5 per cent, compared with up to some 45 per cent for a conventional system. This lower ripple allows a higher mean voltage and hence high corona current to be applied to the precipitator such that an enhanced performance is obtained. Also the use of a three phase supply means that the power factor, referred to the mains, is around 0.94, as against a conventional mains system having a maximum power factor of 0.63, so efficiency savings can be made.

The results of tests on a slip stream precipitator at Plant Miller of Alabama Power showed that SMPS energisation resulted in a 30 per cent decrease in the penetration over a conventional energisation system, with a slightly higher performance being achieved by pulsing the SMPS with a very short pulse width [20]. The results of these tests are indicated in Table 8.2.

Another installation on a paper mill precipitator, where the existing conventional energisation unit was replaced by this SMPS system, received the

Table 8.2 *Test conditions and data – plant Miller tests*

Power supply mode	Date	Test No.	Inlet temp. °F	Boiler load MWe	Flow rate kacfm	Sec voltage kV	Sec. amp. mA	Inlet loading gr dscf^{-1}	Outlet loading gr dscf^{-1}	BHA response outlet	Penetration %
SMPS w/o pulsing (Test 1)	3/3/00	1	380	702	3602	42	10	2.05	0.062	0.175	3.02
SMPS w/o pulsing (Test 2)		2	392	702	3678	42	10	1.95	0.063	0.163	3.23
SMPS w/o pulsing (Test 3)		3	393	700	3738	42	10	1.71	0.077	0.174	4.50
SMPS w/pulsing (Test 1)	3/6/00	1	390	690	3286	30	10–15	2.09	0.042	0.062	2.01
SMPS w/pulsing (Test 2)		2	393	692	3358	30	10–15	1.65	0.043	0.057	2.61
SMPS w/pulsing (Test 3)		3	392	691	3283	30	10–15	1.72	0.047	0.056	2.73
60 Hz SCR (Test 1)	8/25/00	1	400	701	3112	28.7	3.89	2.28	0.069	0.085	3.03
60 Hz SCR (Test 2)		2	394	701	3141	29.1	3.85	1.51	0.068	0.092	4.50
60 Hz SCR (Test 3)		3	391	701	3309	29.7	3.49	2.18	0.100	0.117	4.59

following comments from the site. 'Prior to fitting the SMPS unit, the precipitator operated in the low 30 kV region and was sparking heavily. When the SMPS unit was installed we were able to increase the secondary voltage to approximately 40 kV – a dramatic improvement. The SMPS now runs current limited at or near 40 kV with a clear stack'.

These demonstration tests, although limited, nevertheless indicate that this form of SMPS unit results in benefits similar to those found by other suppliers of SMPS equipment.

The following table provides a comparison of SMPS operational data against a similar rated conventional unit.

Parameter	SMPS	60 Hz T/R
Output voltage kV d.c. average	70	65
Output current mA d.c.	1000	1000
Output power kW	70	65
Peak output voltage kV	71.8	109.2
% Ripple p–p kV	3–5	35–45
Input voltage (V a.c.)	460	460
Input phases	3	1
Input line current line (Amps a.c.)	98.8	237
Losses (kW)	4.0	3.5
Power factor	0.94	0.63
Input kVA	32.2	109
Operating frequency	25 kHz	60 Hz
Arc shutdown time	30 µs	8.33 ms
Cooling	Forced air $\frac{1}{2}$ HP fan	Natural convection
Volume envelope (cubic metres)	0.62	1.70
Plan envelope (square metres)	0.74	1.67
Weight (kg)	340	1750
Litres of fluid	122	613
Wiring from T/R to control	Factory	Field

In summary, based on these findings and consideration of the above comparison with conventional energisation, the supplier considers that the SMPS system offers the advantages to the precipitation industry of a smaller compact unit, being lower in weight, containing less coolant liquid and producing a much improved collection efficiency. The smaller unit enables a change in precipitator configuration and the higher power factor and lower input currents significantly reduces the installed costs.

8.10.5 *Advantages/conclusions reached from these field trials*

- SMPS energisation generally results in significant particulate emission reductions.

- SMPS energisation significantly increases the useful power input into a precipitation field.
- SMPS is a more efficient power converter than a conventional T/R (SMPS *c*. 95 per cent efficiency versus the maximum conventional T/R *c*. 85 per cent efficiency). This improvement in efficiency results in considerable energy savings.
- Readily achieves compliance with 'mains cleanliness codes'. A conventional single phase mains operated T/R disturbs the mains supply significantly and in an unbalanced way. A (three phase) SMPS system complies with such codes automatically because of the way in which it is designed.
- Retrofitting of an ESP with a complete SMPS system, instead of making an upgrade with only new microprocessor controls for a conventional T/R, is more cost effective and the performance advantages of the SMPS approach are complimentary.
- As the SMPS system only contains a minimal coolant volume, no oil tray with expensive drainage pipes and collector tanks is needed, thereby reducing site costs.
- The simplicity of the SMPS installation reduces installation costs and as it can operate through the same communication network as conventional T/R controllers, it is readily connected to the same process and monitoring PCs.
- The replacement of old conventional T/Rs containing PCB oil, with new SMPS systems is a very cost effective and viable alternative, particularly as the performance advantages from the ESP effectively come 'free of charge'.

With the above advantages the SMPS approach will have a significant impact on future precipitator technology and it is suggested that most new applications will have SMPS energisation. Not only should this result in smaller and lower cost installations but, with the ability readily to modify the output waveforms, the impact of reverse ionisation problems should be minimised without the need of equipment additions.

8.11 References

1   www.is.siemens.com/test-facilities/ep/pic410f_e.htm
2   WERNEKINCK, E.: 'A High Frequency a.c./d.c. Converter with Unity Power Factor and Minimum Harmonic Distortion', *IEEE Trans. on Power Electronics, 1991,* **6**, No. 1
3   WIZERANI, M.: 'Direct a.c.-d.c. Matrix Converter Based on Three Phase Voltage Source Converter Modules'. 19th International Conference on Industrial Electronics, 1993, pp. 812–17
4   DEVINE, P., *et al.*: 'A High Power Unity Power Factor Three Phase Rectifier for High Voltage Applications'. Proceedings University Power Engineering Conference, 1996, **2**, pp. 417–20
5   NAKAOKA, M.: 'The State of Art-Phase Shifted ZVS-PWM Series and Parallel Resonant d.c.–d.c. Power Converters Using Internal Parasitic

Circuit Components and New Digital Control', *IEEE Trans. on Power Electronics*, **8**, 1992, pp. 62–70

6   CASADEI, D.: 'Space Vector Control of Matrix Converters with Unity Power Factor and Sinusoidal Input/Output Waveforms'. Proceedings of 5th European Conference on *Power Electronics and Applications*, 1993, pp. 170–75

7   CHO, J. G.: 'Soft Switched Matrix Converter for High Frequency Direct a.c. to a.c. Power Conversion' *Int. Jnl. of Electronics*, **72**, No. 4, pp. 669–80

8   PETKOV, R.: 'Optimum Design of a High Power, High Frequency Transformer', *IEEE Trans on Power Electronics*, 1996, **11**, No. 1, pp. 33–41

9   PEREZ, M. A.: 'A New Topology for High Voltage, High Frequency Transformers'. Proceedings IEEE Applied Power Electronics Conference, 1995, pp. 554–9

10 TALI-IGHIL, B.: 'A New Method to Eliminate the Effects of the High Voltage Transformer Parasitic Elements in a Static Resonant Supply'. Proceedings of 4th European Conference. on *Power Electronics and Applications*, 1991, pp. 468–73

11 DEVINE P. et al. A High Power High Frequency Transformer for High Voltage Applications. Proceedings University Power Engineering Conference. Sept. 1997. Vol. 11. pp 511–514.

12 UK EPSRC Grant No. GR/K 40925: 'A Controllable Variable Waveform High Voltage Power Supply for Electrostatic Precipitators (Leicester University, 1995)

13 RANSTAD, P., and PORLE, K.: 'High Frequency Power Conversion: A New Technique for ESP Energisation'. EPRI/DOE International Conference on *Managing Hazardous and Particulate Air Pollutants*, Ontario, Canada, 1995

14 KIRSTEN, M., *et al.*: 'Advanced Switched Integrated Rectifiers for ESP Energisation'. Proceedings 8th ICESP Conference, Birmingham, Al., USA, 2001, Paper C1-3

15 SRINAVASACHAR, S., *et al.*: 'Ultra High Efficiency ESP for Fine Particulate and Air Toxic Control'. DoE Contractors Meeting, Pittsburg, Pa., USA, 1997

16 REYES, V., *et al.*: 'A Novel and Versatile Switched Mode Power Supply for ESPs'. Proceedings 7th ICESP Conference, Kyongju, Korea, 1998, pp. 339–51

17 REYES, V., and LUND, C. R.: 'Full Scale Test with Switch Mode Power Supplies on an ESP at High Resistivity Operating Conditions'. Proceedings 8th ICESP Conference, Birmingham, Al., USA, 2001, **II**, Paper C1-4

18 WETTELAND, O.: 'Hard Switching – A Superior Switched Mode Power Supply Design for ESP'. Proceedings 8th ICESP Conference, Birmingham, Al., USA, 2001, **II**, Paper C1-2

19 MOHAN, N, *et al.*: 'Power Electronics: Converters, Applications and Design' (John Wiley and Sons, 2nd Edition, 1995, pp. 249–97

20 SEITZ, D., and HERDER, H.: 'Switch Mode Power Supplies for Electrostatic Precipitators'. Proceedings 8th ICESP Conference, Birmingham, Al., USA, 2001, **II**, Paper C1-1

Chapter 9

The impact of electrical resistivity on precipitator performance and operating conditions

As indicated in Chapter 3, the electrical resistivity and cohesivity of the particulates presented to the precipitator play an important role in its electrical operation and hence performance. This chapter will examine how these factors impact on the operation of precipitators and the non-electrical means of mitigating the impact of these fly ashes on performance, the electrical methods being described in Chapter 7.

9.1 Particle composition

In general, the actual chemical composition of the particulates does not have a significant effect on the mechanical design of a precipitator, unless it is chemically reactive, when corrosion resistant materials must be used for fabrication purposes.

Most dusts from industrial processes consist of a large number of chemical constituents in differing particle sizes, which arise from feed stock impurities or the fuel residues used in the actual processing. In the case of metallurgical processes, the particulate matter usually consists of a high proportion of metallic oxides, which are initially volatilised in the high temperature part of the process. The materials that have been volatilised usually condense out of the waste gas stream in the cooler temperature regions of the plant, typically following heat recovery. Particles formed by condensation are usually spherical and of small size, typically less than 1 µm diameter. When designing a precipitator to handle this type of particulate one must be aware of the penetration window in the fractional efficiency curve (see Figure 1.9), and also the possible impact of the fine fume in terms of corona suppression due to their presence in the inter-electrode area. This corona suppression aspect may require the use of high emission discharge electrodes as discussed in Section 4.2.

For coal and other combustion processes, although about 1 per cent of the ash components can be volatilised within the furnace, the temperatures are generally too low to fuse silica and alumina compounds completely; these are usually present in the waste gases as irregular shaped particles, resulting from comminution of the fuel. Dependent on impurities present in the coal ash, some of the silica/alumina can combine at the high furnace temperatures with these impurities to produce a eutectic, which can be found in the form of glassy spherical shaped particles. If carbon becomes trapped in the spherical particles, 'cenospheres', i.e. hollow spheres, can be produced, which have a large surface area but little mass, so are subject to scouring and reentrainment from the collectors. These cenospheres do not impact on electrical operation, but generally because of their low mass and poor cohesivity can be readily re-entrained, thereby detracting from overall performance.

9.2 Particle resistivity

For steam raising units, the coal ash carried forward to the precipitator normally comprises a high percentage of silica and alumina, both excellent insulators in their own right having very high resistivity characteristics; whereas those that tend to reduce the resistivity in 'cold side' applications sited downstream of the air heater are the sulphur in the coal and sodium oxide in the ash. For so-termed 'hot side' units, situated upstream of the air heater and operating at a temperature of some 300 °C, the charge transference mechanism is different, relying more on improved thermal transference properties within the particle matrix, resulting from the increase in kinetic energy at the higher temperature.

The electrical resistivity of the particulate is determined by two mechanisms: the bulk volume conductivity, which is a function of the particle matrix constituents, and by surface conduction, which is controlled by the particle surface reactivity and gas components. Figure 9.1 indicates the difference in conduction paths.

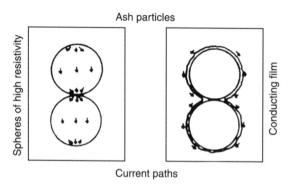

Figure 9.1   Comparison of particulate charge conduction paths

9.3 Measurement of particulate resistivity

The measurement of resistivity, because of this surface conditioning effect, makes laboratory determination difficult. There are, however, standard techniques and apparatus where the sample is placed in a cell and the surrounding environment simulates the original gas condition in terms of temperature, water dew point and, if necessary, acid gas concentration. To obtain comparative laboratory data, the accepted procedure uses a standard cell configuration and the dust is compacted to a set pressure [1]. Once the apparatus has reached a predetermined 'equilibrium gas' condition and temperature the current passing through the cell is determined for a range of applied voltages. In practice, resistivity evaluations are made for both increasing and decreasing temperatures to enable the peak and general resistivity profiles to be determined.

These laboratory measurements tend only to be comparative, because the packing density will not necessarily be the same as on the collector plates and the sample itself may have aged and so affected its surface reactivity. In addition, measurements by Goard and Potter [2] have shown that the resistivity, not unexpectedly because of the ionic nature of the conduction mechanism, is strongly field dependent.

For *in situ* field determinations, the US Southern Research Institute (SoRI) developed a point/plane apparatus [3], which is inserted into the flue. Dust is electrostatically precipitated onto the plate by energising the point electrode negatively, with respect to the earthed plane or plate electrode. After a certain time, depending on the dust concentration, a circular plate, concentric with the discharge point, is carefully lowered onto the dust surface and the current, for a preset range of voltages, is measured. From this, and with known cell dimensions, the resistivity of the precipitated dust can be evaluated.

An alternative form of apparatus for both *in situ* and laboratory measurements is a cell mounted in the base of a small 25 mm diameter sampling cyclone [4]. Gas is drawn through the cyclone, which collects all but the smaller submicrometre particles (99 per cent efficiency at 1 μm diameter), and after rapping the cell to produce a constant packing density, the current through the cell is measured in the usual way, thereby enabling the resistivity to be calculated.

With either type of *in situ* apparatus, although the measurements are carried out under real carrier gas/environmental conditions, the actual ash sample is extracted from only a single position in the ducting and, therefore, may not be truly representative of the total ash. There is also the possibility that any 'fines' may escape the apparatus completely, so due care must be exercised in analysing the resultant data.

A typical laboratory derived temperature/resistivity curve for a power station fly ash is illustrated in Figure 9.2. This shows that at temperatures above some 200 °C, where any possible surface conditioning effects may have been destroyed, the resistivity follows the classical theory of volume conduction, as a result of increased thermal motion of the molecules, thereby producing an approximate linear relationship with the inverse of absolute temperature. Below

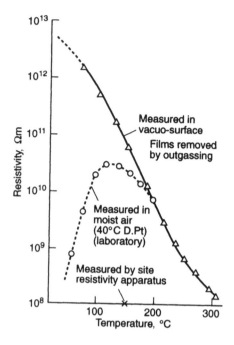

Figure 9.2   Laboratory resistivity of power station fly ash

160 °C, the curve shows a decreasing resistivity as the surface conditioning, in this case owing to moisture deposition, becomes more significant as the temperature falls as a result of the increasing vapour pressure.

It was possible, because of the laboratory situation in this instance, to subject the system to vacuum, which effectively removed all gaseous phase components from the equipment. Under this situation even at low temperature the fly ash resistivity steadily increased because of the reduction in kinetic energy of the molecules. The results confirming that, at high temperature, the fly ash resistivity is controlled by the composition of the particle matrix and charge transference is through the matrix itself, whereas, at low temperature, conduction is mainly dictated through a surface conditioning layer, as confirmed by the outgassing experiment.

Although many reverse ionisation experiences are related to coal fired power station precipitators, there are other processes that, unless suitably conditioned, are also subject to reverse ionisation phenomena. Figure 9.3 has been developed for a dry process cement kiln dust that illustrates the effect of changing gas moisture content on particulate resistivity. From this, one can appreciate the need for water conditioning of this type of dust in order to avoid the problems of reverse ionisation and therefore to limit the size of the precipitator. Many metallurgical processes also employ moisture conditioning to cool the gases in order to minimise any potential resistivity effects in the precipitator.

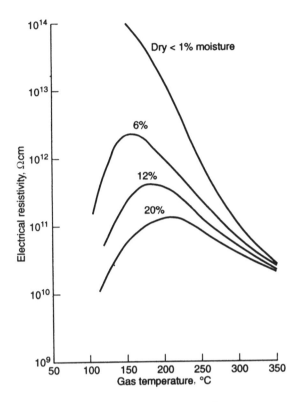

Figure 9.3   Laboratory resistivity of dry process cement dust

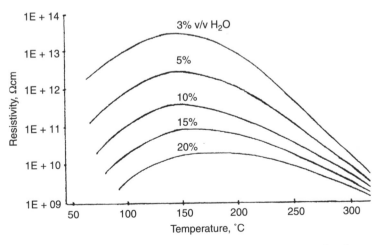

Figure 9.4   Fly ash electrical resistivity change with gas temperature and moisture content

Figure 9.4 indicates typical fly ash resistivities for different gas moisture contents over a range of temperatures. These curves were produced from laboratory measurements using a fly ash from a medium sulphur content coal, which had been collected from an operating precipitator.

At the higher temperatures the curves tend to come together irrespective of the moisture content of the gas, whereas at the lower temperatures the effect of gas moisture is very marked. These data supporting the difference in charge conduction mechanisms are dependent on the temperature and gas equilibrium conditions.

Although many engineers consider the electrical resistivity of the particulates to be all important, the problems of accurately determining the true value, as indicated above, really means that resistivity can be only one of the tools used by the precipitation industry for the design and sizing of precipitators.

9.4 Resistivity effects on low temperature power station precipitators

During combustion, most of the coal's pyritic sulphur burns to produce sulphur dioxide, although a small percentage converts to sulphur trioxide, which is important on 'cold side' applications to reduce the fly ash electrical resistivity. In a boiler waste flue gas there is generally at least 6 per cent moisture present, either from the moisture or converted hydrogen in the coal plus atmospheric moisture in the combustion air. At the equivalent dew point temperature, any free gaseous SO_3 reacts with this gas phase moisture to produce sulphuric acid mist, which then uses the particles as condensation nuclei. The subsequent thin/monatomic layer of sulphuric acid deposited on the fly ash enables charge transference to occur through this high conductivity layer.

The actual effect of surface conditioning in this case is dependent on both the amount of sulphur trioxide and moisture present, plus the gas temperature, which dictates the saturation vapour pressure. The effect of sulphur trioxide concentration in the flue gas upon the acid dew point temperature for a moisture content of 6 per cent is indicated in Figure 9.5.

For coals containing in excess of 1 per cent sulphur, there is generally sufficient natural surface conditioning at normal operating temperatures of around 130 °C to give acceptable values of resistivity for effective precipitation. Although this is generally true, there are incidents where the fuel, particularly if the hydrocarbon content is not totally combusted, is initially volatilised and later condenses on the fly ash particles as an insulating type of coating that can promote a severe reverse ionisation type of operation. This phenomenon/occurrence is probably more prevalent during plant start up, when the furnace is cold and the oil start up burners are not ideally set up; however, the situation can arise at other times, so one must be aware of this potential problem. In some start up instances, although the initial emission from the stack might be high, it reduces with time as combustion improves and the initial deposits of any

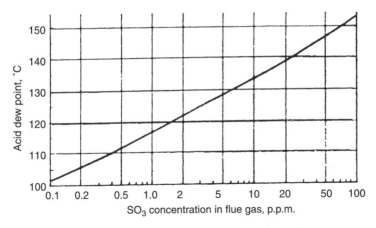

Figure 9.5   Effect of sulphur trioxide concentration on acid dew point temperature

contaminated fly ash are removed from the collectors. In the case of a really fouled collector system some manual cleaning may be necessary to remove the deposits to regain a satisfactory precipitator performance.

The presence of sodium in the ash behaves in a very different manner where, instead of surface conditioning, the sodium ions within the particle matrix act as charge carriers, which assist in negating the effect of high resistivity fly ash to a certain extent. Measurements on a Japanese installation firing Australian low sulphur (0.5 per cent) fuels, and therefore expected to give 'difficult' precipitation, showed that the precipitator emission was very dependent on the sodium oxide content of the fly ash, as illustrated in Figure 9.6 [5]. Lithium is also anticipated to react similarly, but as it is usually present in extremely small quantities, it does not have a significant impact on performance. Potassium, although having similar chemical properties to sodium, does not appear to react in the same manner.

Calcium and magnesium present in the ash tend to produce sulphates with their exposure and reaction with sulphur trioxide in the gas stream. These sulphates are chemically produced and, having low conductivity, tend to interfere with the potential resistivity reduction by sulphur trioxide condensing as sulphuric acid mist onto the fly ash. Calcium and magnesium components present in the fly ash therefore must be considered as leading to lower precipitator performance through increased resistivity effects and appropriate adjustments made in the design/sizing of any installation. The resultant ash from certain lignite and sub-bituminous coals can have lime contents of around 20 per cent, which have a significant impact on the initial precipitator sizing.

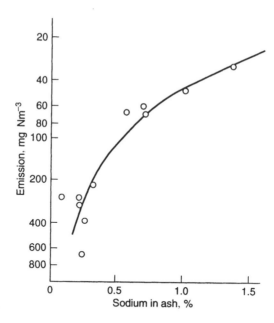

Figure 9.6   The effect of sodium oxide in the ash upon efficiency

9.5 Correlation between precipitator performance and particulate resistivity

Performance measurements on a large number of power plant precipitators have been correlated with the coal and fly ash analyses, enabling first order precipitator sizing curves to be developed. Figure 9.7 indicates the Deutsch migration velocity necessary for a precipitator to achieve a nominal efficiency of 99.5 per cent, under normalised conditions of temperature, dust loading, etc., against the coal sulphur plus ash sodium oxide contents [5]. Other curves, incorporating silica, alumina, calcium and magnesium, in addition to coal sulphur and sodium oxide in the ash, have been developed but these do not appear significantly to affect precipitator sizing requirements based on data from Figure 9.7.

   Although the gas components, ash chemistry and reactivity at the precipitator inlet are important in that they determine the particle resistivity and hence performance of the installed precipitation plant, at the design stage of most proposals the only source of information available is the raw coal and ash analyses, together with possibly operating temperatures at the inlet to the precipitator. These analyses, however, can provide the precipitator design engineer with invaluable information on the probable fly ash resistivity and hence the potential size of precipitator for a specific efficiency. In most new applications the fly ash electrical resistivity is computed from the coal and ash chemistry by using a relationship developed by Bickelhaupt [6]. This has proved useful to the Precipitation Engineer, because one can assess what changes can be theoretically

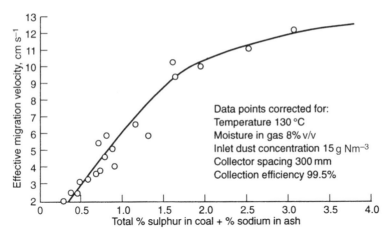

Figure 9.7   Effect of sulphur in coal plus sodium in ash upon performance

applied to the inlet gas conditions and/or fly ash chemistry in order to modify the resistivity. The results can then be assessed as to the practicality of the approach, whether it is by cooling the gas or changing the surface effect by flue gas conditioning techniques.

Measurements of *in situ* fly ash resistivity (Hall [7]) show acceptable agreement with those calculated from the Bickelhaupt relation. Figure 9.8 is a plot showing

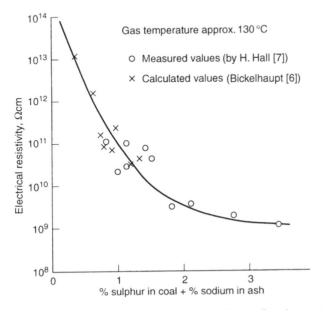

Figure 9.8   The effect of sulphur in coal plus sodium in ash upon fly ash resistivity

how the resistivity varies predominantly with the sulphur in the coal plus sodium oxide contents of the ash. From this, it is interesting to note that, at around a resistivity 2×10^{10} Ωcm, the resistivity curve steeply increases with reducing values of sulphur plus sodium, whereas for higher values of sulphur plus sodium oxide the resistivity decreases only slowly. The resistivity level of 2×10^{10} Ωcm is that proposed by White [8] as being a critical value of resistivity between good and poor precipitator performance. This resistivity relationship, shown in Figure 9.8, basically mirrors the shape of the performance curve shown in Figure 9.7.

For 'cold side' power plant precipitator applications one of the methods adopted for reducing the fly ash resistivity to overcome reverse ionisation effects is by the injection of small quantities of gaseous sulphur trioxide, around 15 p.p.m. v/v, into the gas stream upstream of the precipitator inlet. Upon injection, the sulphur trioxide reacts with the moisture in the gas stream to form a sulphuric acid mist, which then uses the fly ash as condensation nuclei to produce a monatomic surface layer of conducting acid, which reduces the electrical resistivity of the particles. Figure 9.9 shows a CEGB photomicrograph of unconditioned and sulphur trioxide conditioned fly ash obtained during initial conditioning work in the UK, indicating the extremely thin deposited layer of acid [9].

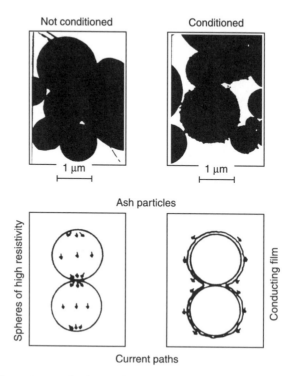

Figure 9.9   Photomicrograph of unconditioned and conditioned fly ash particles

Although other reagents, such as phosphorus, inorganic salts containing sulphate radicals, and organic materials, have been investigated, the largest number of flue gas conditioning systems in commercial use are based on sulphur trioxide. As the conditioning effect is condensation of acid mist onto the particles it is important that, prior to injection into the gas stream, the sulphur trioxide remains above the acid dew point temperature. The quantity of reagent required to modify the resistivity is dependent on the flue gas temperature, rising sharply as the temperature approaches 160 °C. Above 160 °C, the effect of sulphur trioxide injection diminishes and alternative approaches need to be considered.

For 'hot side' power plant applications at a temperature of around 300 °C, although the fly ash resistivity is usually less than 10^{12} Ωcm (see Figure 9.2) and hence operation should be free from reverse ionisation effects. In a large number of US installations where 'hot side' applications once prevailed, a similar but specific electrical phenomenon occurred in that electrical operating conditions tended to mirror 'cold side' reverse ionisation. Investigations found that the release of positive ions from the deposited ash was related to the sodium ions contained in the ash actually migrating toward the collector plate [10]. This resulted in a sodium depleted layer of high resistivity ash being produced, which reduced the rate of particle charge neutralisation, allowing voltage to build up on the surface similar to the 'cold side' phenomenon. In order to overcome this situation, some plants have opted for the injection of sodium based materials to increase the quantity of sodium ions available for charge transfer. Some other installations have modified the ductwork arrangement such that the precipitator is now repositioned downstream of the air heater and advantage is taken of the reduction in gas volume with temperature using the existing precipitator sizing to increase the overall performance.

Although any sodium based material can be applied, the cost, availability and security of supply usually limits the choice to sodium carbonate, sulphate or nitrate. Various conditioning systems are available, such as by adding the material to the coal feed, injection of a dry powder in the furnace area or the evaporation of injected liquor into the gas stream. Whereas the simplest and easiest approach for demonstration purposes is possibly by adding the product to the coal feeders, the sodium in the high temperature zone can, however, react with other ash compounds to form a corrosive slag, which can damage the furnace wall and superheater dependent on the temperatures and ash composition. A further problem with adding sodium salts to the coal feeding system is that of possible high corrosion rates on the coal feeder and milling equipment.

An alternative conditioning system to mitigate the corrosion risks is to inject a concentrated liquor containing sodium compounds directly into the gas stream at some convenient location where the carrier liquor will be evaporated. This is a more complicated approach, however, requiring storage tanks for the liquor, control and dosing pumps plus a liquid distribution system and possibly dual fluid air atomising sprays to ensure evaporation and elimination of liquid carryover into the downstream areas of the plant.

With some Australian 0.5 per cent sulphur coals, which can produce an ash comprising some 90 per cent silica plus alumina complex, precipitation difficulties arose; not only was the ash highly resistive, resulting in reverse ionisation, but the cohesivity of the deposited ash was low, which caused severe re-entrainment problems, both situations adversely impacting on the overall performance. In this instance, the difficulties were overcome by dual flue gas conditioning using simultaneous injection of sulphur trioxide to reduce the resistivity and anhydrous ammonia to increase the cohesivity, both chemicals being injected at rates of around 15 p.p.m. v/v.

9.6 Low resistivity ash and its effect on performance

In the power industry it is fairly usual to consider carbon carryover as approximately equal to the loss on ignition at 800 °C. In other industries, however, the loss on ignition can be the result of the loss of water, the release of carbon dioxide from carbonates, or the evaporation of alkali salts and some metallic components, so for industries outside power generation, the carbon needs to be determined in some other way.

In the case of power plants, carbon is usually present in the form of partially burnt coal, comprising large particles of low mass, typically having a bulk resistivity of around 10^8 Ωcm. Although the carbon particles can be satisfactorily charged by the corona, as they approach the collectors they lose their charge so rapidly as to be repelled back into the gas stream where they can be re-entrained. This charging and repulsion situation can occur numerous times as the particle passes through the precipitator and some particles can eventually escape without being collected. Even if the particles are retained on the collectors, the fly ash plus carbon deposit has very low cohesivity and upon rapping tends to shatter/explode, rather than being sheared off the collector, so can be readily re-entrained by the horizontal gas flow. In either situation the efficiency of collection will be reduced.

To overcome the low cohesivity effect, and to some extent the charge/repulsion phenomena, flue gas conditioning using ammonia can be applied in a similar manner to the sulphur trioxide system. A typical method of conditioning is to evaporate anhydrous liquefied ammonia. This is then mixed with dilution air to produce a nominal ammonia injection rate of around 10–15 p.p.m. v/v. Upon injection, the gaseous ammonia reacts with the sulphur trioxide and dioxide in the flue gases to produce ammonium sulphate/sulphite compounds. Some of the compounds can be sticky at the flue gas temperature, which upon collection increases the overall cohesivity of the deposit, such that rapping re-entrainment is reduced.

Although other reagents such as ammoniacal liquor and ammonium salts can be used, it is important to control the injection rate such as not to overcondition and produce ammonium bisulphite which can condense as the gases emerge from the chimney into a cooler temperature region. This situation is readily noticeable as a whitish discharge in the form of a sub-micrometre fume.

While the larger carbon particles are deleterious to precipitator performance, if the carbon is in the form of fine particles, such as carbon black, their lower saturation charge is insufficient to result in their being repelled back into the gas stream and they are collected with the other materials. The mixture of fine carbon plus the other materials can result in a significant reduction in the bulk resistivity of the deposit, which aids performance.

Working with low resistivity spray drier particles, Durham [11] showed that, by theoretically examining the conditions existing at the boundary of the dust layer (Figure 9.10), it is possible for a repulsive electric force to develop that expels the particles back into the gas stream.

From Figure 9.10, an electric field E_g exists at the surface of the layer as a result of the corona electrode voltage, which also creates the ion density N_i at the surface.

The resultant current density J_g is in the positive x direction (although the ions move toward the collector), is given by:

$$J_g = N_i \times e \times b_i \times E_g \text{ A m}^{-2}, \tag{9.1}$$

where N_i is the ion density, e is the charge on an electron and b_i is the ion mobility.

For homogeneous and steady state conditions

$$J_g = J_i = J_p \text{ A m}^{-2}, \tag{9.2}$$

where J_i is the current density through the dust and J_p is the specific collector current. Hence

$$E_i = J_i \times \rho_1 \text{ } V \text{ m}^{-1} \tag{9.3}$$

where E_i is the electric field within the dust layer and ρ_1 the dust resistivity (Ωm).

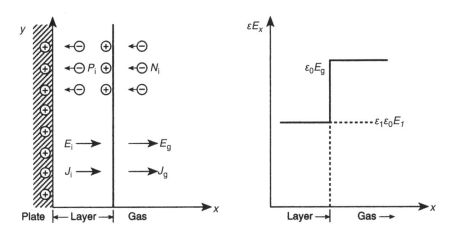

Figure 9.10   Forces acting on the dust boundary

The surface charge density is given by

$$\sigma = \varepsilon_0 \, (E_g - \varepsilon_i \times \rho_1 \times J_p) \; \text{c m}^{-1}, \tag{9.4}$$

where ε_0 is the dielectric constant for free space and ε_1 is the dust layer dielectric constant. From this, the surface charge density can be positive or negative, depending on the resistivity and layer conditions.

The force acting on the surface charge density, f_x, is given by

$$f_x = (E_g + E_1)/2 \quad \text{N m}^{-2}. \tag{9.5}$$

If the surface charge density is positive, the force tends to pull the particles from the surface, and if negative, the force holds the particles to the surface.

Although theory would suggest that low resistivity dusts should be subject to electrical re-entrainment (repulsion), there are a number of full-scale plants fitted with precipitators following spray driers used for desulphurisation, where re-entrainment does not appear to be a problem. This does not necessarily mean that the proposed theory is incorrect but, as the properties of the particulates are controlled by a large number of factors, it is likely that a slight difference in the approach to dew-point temperature could make the dust more adhesive in spite of the lower resistivity.

9.7 Equipment for flue gas conditioning

Although the theory of flue gas conditioning is readily understood, the equipment for injecting the conditioning agents is quite complex. There are three major equipment suppliers, which offer slightly different types of equipment, each having preferred or patented approaches. These differences are basically concerned with improving the overall plant efficiency in terms of feedstock usage and operating power, which tend to be supplier and site specific.

To minimise site installation costs, all equipment is largely skid mounted, pre-wired and checked prior to delivery. The site work is thus limited to the final injection manifold and distribution pipe-work, electrical hook up and inter-facing with the plant control system. The basic approaches, however, to produce the conditioning agents are similar, in that for SO_3 injection, the system converts sulphur dioxide to trioxide by means of a catalyst before dispersion into the ductwork, while for NH_3 injection schemes, evaporated gaseous ammonia is mixed with air before dispersion.

9.7.1 Sulphur trioxide conditioning

In the case of sulphur trioxide conditioning, the feedstock can be liquefied sulphur dioxide, which is stored and evaporated adjacent to the plant, or liquid or molten sulphur, which is contained in an adjacent holding tank, while the latest approach is to use solid sulphur, which is used similarly to a pulverised fuel. For a sulphur feed stock, the sulphur is fired in a sulphur burner with a balanced amount of air to produce around a 5 per cent concentration of SO_2. This is then passed through a vanadium pentoxide catalyst chamber, where the sulphur dioxide is converted to sulphur trioxide with an efficiency of some 95 per cent. Using staged catalyst beds with inter-stage cooling because of the exothermic reactions, conversion efficiencies greater than 95 per cent can readily be achieved. The sulphur trioxide exiting the catalyst bed is then diluted with hot air and passed through a distribution manifold into the ducting upstream of the precipitator. A typical flue gas conditioning system, using staged catalytic conversion to produce the sulphur trioxide, is illustrated in Figure 9.11 [12].

To maximise the conditioning effect, in a contact time of <1 s, it is important that the reagent gases are introduced into the flue gas having as wide a contacting area as possible. A typical injection manifold would comprise a number of probes spaced across the duct, each having injection nozzles positioned at the 'centres of equal areas'. To avoid corrosion and pluggage problems within the gas distribution network, the manifold system is constantly maintained at a temperature above any potential acid dew point.

Investigations to convert the naturally occurring sulphur dioxide in the flue gases on power plants is being studied as an alternative to imported feed stock conditioning. With these approaches, a parallel side stream catalyst bed is installed upstream of the economiser, where the temperature is around 400 °C, and the raw flue gas passed through the catalyst to convert some of the dioxide to trioxide [13, 14]. Two systems are under consideration, one where the gas flow through the bed is adjusted to produce the correct amount of sulphur trioxide for conditioning, the other, where the area of the catalyst is varied with a constant gas flow; again the object is to convert only the correct amount for fly ash

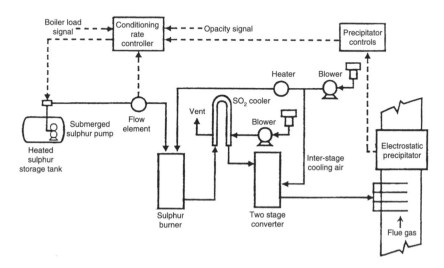

Figure 9.11   Typical sulphur trioxide injection equipment using staged catalytic conversion

conditioning, otherwise serious corrosion and the discharge of acid mist from the chimney could occur. These latter approaches, although avoiding the need to import feedstock material, not only impact on the overall boiler efficiency, but can also prove very expensive in that unless pre-cleaning of the gases is adopted, the cost of new/replacement catalyst beds can be significant. There is also the catalyst ageing factor to be considered, which will reduce the nominal design conversion efficiencies and hence produce potential precipitator performance shortfalls.

Figure 9.12 illustrates a summary of the various sulphur trioxide flue gas conditioning technologies that can be used dependent on the preferences and specifics of the customer [15].

9.7.2 Ammonia conditioning

The equipment for ammonia injection is simpler in that anhydrous liquefied ammonia is stored and evaporated adjacent to the plant and mixed with air before passing through a distribution pipe manifold system. A typical ammonia injection system is illustrated in Figure 9.13. Although 25 per cent ammoniacal liquor, a by-product from a coke oven works, can be used, this requires evaporation prior to dilution. The availability of ammoniacal liquor to the site may mean that this approach is not viable in spite of a much lower cost and easier handling.

As previously indicated, with some particulates, e.g. Australian high silica plus alumina fly ashes, dual conditioning is required. This is achieved by having both a sulphur trioxide and ammonia injection system operating in parallel, but injecting the ammonia upstream of the sulphur trioxide. This enables the ammonia initially to react with the gaseous sulphur dioxide to produce ammonium bisulphate/bisulphite to enhance the particulate cohesivity; the sulphur trioxide is then able to reduce the fly ash resistivity to improve precipitator performance electrically

The actual flue gas injection rates, usually controlled by microprocessor devices, can be up to some 20 p.p.m. v/v for either or both reagents. These tend to be site dependent and are controlled from feedback data from the boiler loading, continuous emissions monitoring system and precipitator electrical operation.

The effect of flue gas conditioning in reducing precipitator emissions from plants handling difficult fly ashes, whether they arise from high electrical resistivity or poor cohesive properties, is illustrated in Table 9.1 [16].

The reductions in emission shown in Table 9.1 are based on 6 min average opacity readings or actual site determinations.

The tabulated data indicate that, dependent on the application, flue gas conditioning can bring about a minimum emission reduction of 50 per cent and, in certain circumstances, can approach nearly 90 per cent.

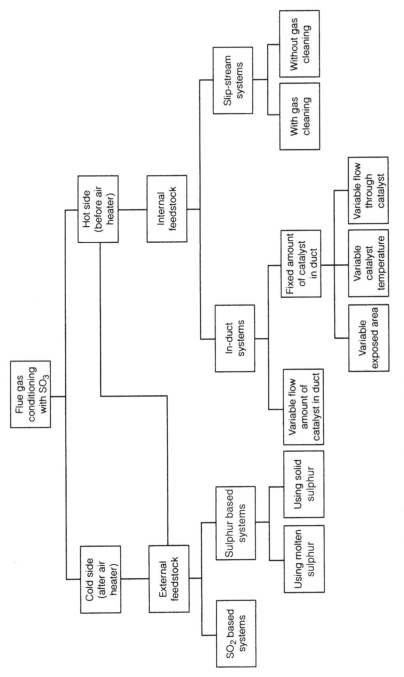

Figure 9.12   Summary of SO₃ flue gas conditioning technologies

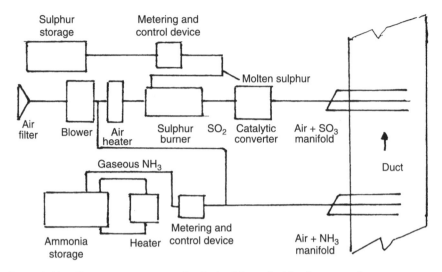

Figure 9.13   Generic arrangement of a dual sulphur trioxide plus ammonia conditioning system

Table 9.1   Emission reductions achieved by flue gas conditioning

Fuel type	Conditioning agent, p.p.m. v/v SO_3	NH_3	Emission reduction percentage
USA low sulphur Eastern	12	6	74
USA low sulphur Western	8	0	>75
Spanish lignite + South African Bit.	0	15	87
South African 0.7% sulphur	20	0	81
Australian 0.5% sulphur	10	0	87
Polish 0.7% sulphur	20	0	57
Indian high ash low sulphur	20	0	50
Indian high ash low sulphur	0	15	80
Australian 0.5% sulphur, high $Al_2O_3 + SiO_2$	0	15	80 + IE*

* IE = intermittent energisation

9.8 Humidity conditioning

As indicated earlier, the moisture content of the flue gases has an important role in determining the electrical resistivity of the particulates and hence precipitator electrics and performance. Although chemical reagent injection is usually considered the preserve of the power industry, it has been used on other process plants to enhance the performance of the precipitation plant. There are, however, a number of processes where moisture conditioning is essential to minimise the precipitator size to produce a cost effective solution. Usually these processes derive their waste gases without the use of carbonaceous fuels, such as electric smelting or oxygen refining of iron, but there are some such as dry process cement manufacture, particularly when the raw meal mill is not in service, which require moisture conditioning to produce acceptable efficiencies.

Generally these processes have specially designed cooling towers preceding the precipitators, where water in injected in fine droplet form to ensure evaporation to avoid carryover problems arising. In addition to reducing the particulate resistivity, moisture injection also decreases the gas volume and temperature of the gases being treated, which reduces the size and hence cost of the precipitator, although a cooling tower will add to both capital and operating costs.

As an alternative to chemical conditioning, on some power plants moisture has been injected into the ductwork immediately in front of the precipitators and achieved significant improvements in performance. The difficulty with this approach is that, with an exposure time of less than 1 s and a gas temperature of around 130 °C, the size of the injected water droplet is critical. To minimise the injected droplet size, two fluid atomisers are essential to ensure complete evaporation of the water in order to eliminate fall out and build up problems downstream of the injection point [17].

9.9 Reducing inlet gas temperature

Rather than injecting moisture to lower the gas temperature to reduce the electrical resistivity, work in Japan employing a gas/gas heat exchanger in front of the precipitator to increase the relative humidity and thus reduce the electrical resistivity of the particles has been shown to be effective. This approach is now a fairly standard practice used by a number of Japanese coal fired installations using difficult low sulphur fuels. The effect of reducing the gas temperature on particle resistivity for a range of low sulphur coals is shown in Figure 9.14 [18].

9.10 Summary of resistivity effects on precipitator performance

The above methods of overcoming the difficulties of a highly resistive fly ash are based on changing the properties of the gas or particulate at the inlet to the precipitator, whether it is upstream or downstream of the air heater. There are, however, alternative methods of mitigating the effect of reverse ionisation on performance, which are of an electrical nature. These methods include intermittent energisation, where instead of a continuous energising voltage, the supply is intermittent, thereby allowing any charge arriving at the collector sufficient time during the non-energising period to neutralise before the next arrival of corona following re-energisation. Another approach is to superimpose a high voltage ~60 kV, short duration 100 µs, pulse onto a reduced energising voltage. Again the mechanism is that during the pulse period sufficient corona is produced to

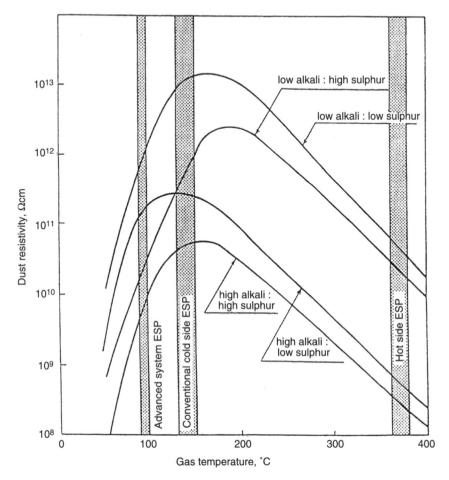

Figure 9.14   Effect of reducing temperature on electrical resistivity for a range of coals

charge the particles and that there is sufficient time between successive pulses for the particle charge to neutralise. Thus with either approach, voltage build up on the fly ash surface is avoided, which would have given rise to a positive ion emission, i.e. reverse ionisation operation.

The effect of resistivity on precipitator operation and performance can be summarised as follows.

(a) For particles having resistivity in the range 10^9–10^{11} Ωcm, the particle charging and discharging regime when the particle arrives at the collector proceeds normally and so has minimum impact on performance (see curve A in Figure 9.15).

(b) As the resistivity increases, although the charging occurs normally, the particle loses its charge only slowly on arriving at the collector, and a voltage

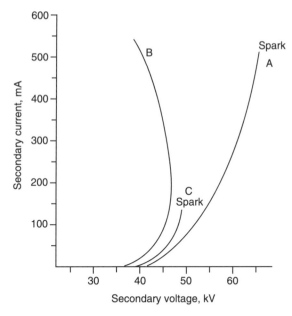

Figure 9.15 V/I characteristics for the three fly ash resistivities

begins to develop across the dust layer according to Ohm's law. Dependent on the resistivity and layer thickness, the voltage, for resistivities in the 10^{13} Ωcm range, can reach a point where positive ions begin to be emitted from the surface of the dust. These positive ions cross the inter-electrode space and collide with and neutralise negative ions and charged particles, which significantly reduces precipitator performance. This condition is termed reverse ionisation, or back-corona, and the operating electrics exhibit a much reduced average voltage but an extremely high current (see curve B in Figure 9.15).

Recent investigations of the applied voltage and current waveforms have shown that, under reverse ionisation conditions, the peak applied voltage tends to be maintained, but with the high current flow, the ripple voltage significantly increases and the main feature is that the minimum ripple voltage level dramatically falls. To control/minimise the deleterious effects, modern AVC units monitor this minimum voltage condition and take appropriate corrective action to prevent the positive current 'runaway'.

(c) For slightly lower resistivities, or dusts having poor packing density, although a voltage develops across the dust layer, interstitial breakdown occurs through the layer and gives rise to a 'streamer', which results in field breakdown. The symptoms of this condition are a slightly reduced average operating voltage, but a very low discharge current, any attempt to raise the current immediately resulting in increased flashover. Examination of the voltage waveform shows a minimum ripple voltage and any attempt to increase the current raises the threshold voltage resulting in breakdown. Although this condition does not have such an impact on performance, as does the classical reverse ionisation or back-corona scenario, it nevertheless causes an efficiency limitation (see curve C in Figure 9.15).

The impact on the electrical characteristics of a free running precipitator is indicated in Figure 9.15, for the above electrical resistivities.

(d) At the lower end of the resistivity range, for very conductive particles such as metallics or carbon, the charge on arriving at the collector is lost so quickly that the particle either sits on the surface as a neutral particle that can be readily re-entrained by the gas stream, or is transferred so rapidly as to repel the particle back into the gas stream. This re-entrainment/repulsion of the particles can occur several times during the particles' passage through the precipitator and can result in particles escaping with the outlet gases. While there are no specific characteristics identifying this phenomenon, the potential re-entrainment results in poor collecting efficiency.

The extent of the re-entrainment depends on a number of factors, such as particle cohesion, gas stream velocity and turbulence, and the electrostatic forces acting on the particle. Examination of outlet dust samples from plants dealing with low resistivity particulates, e.g. carbon, shows an increasing quantity of these conductive particles. For pulverised coal fired units having low NO_x burners producing a carbon carryover of around 10 per cent, the increase in carbon in the outlet gases can be at least 30 per cent, while for carbon carryovers in the 15–20 per cent range, the carbon content on the outlet dusts can be as high as 90 per cent. Although these low resistivity particles generally have no significant effect on electrical operating conditions, the emission from the precipitator is high as a direct result of particle re-entrainment from the collectors.

9.11 References

1 IEEE Std 548: IEEE Standard Criteria and Guidelines for the Laboratory Measurement and Reporting of Fly Ash Resistivity, 1984

2 GOARD, P. R. C., and POTTER, E. C.: 'Operational Resistivity Measurements on Freshly Generated Fly Ashes'. CSIRO Symposium on *Electrostatic Precipitation*, Leura, 1978, pp. 3.1–8; CSIRO, Sydney, Australia

3 NICHOLS, G., and SPENCER, H.: 'Test Methods and Apparatus for Conducting Resistivity Measurements'. Report prepared by the Southern Research Institute for the US, EPA, 1975, Southern Research Institute, Birmingham, Al., USA

4 COHEN, L., and DICKENSON, R.: 'The Measurement of the Resistivity of Power Station Flue Dust', *Jnl. Scientific Instruments*, 1963, **40**, p. 72–5

5 DARBY, K., *et al*.: 'The Influence of Sodium in Fly Ash on Electrical Resistivity and its Impact on Precipitator Performance'. Proceedings EPRI/EPA 9th Particulate Control Symposium, Williamsburg, USA, 1991, Session 6A; EPRI TR 100471 2, Palo Alto, Ca., USA, 1992

6 BICKELHAUPT, R. E.: 'A Technique for Predicting Ash Resistivity'. EPA Report No. 600/7-79-204 US, 1979

7 HALL, H. J.: 'Fly ash chemistry indices for resistivity and effects on precipitator design and performance.' Proceedings Fourth Symposium on the transfer and utilisation of particulate control technology, Vol. II, Electrostatic precipitation, pp. 459–473. Shamrock Hilton, Houston, TX, USA, 1984. EPA 600-9-84-0256.

8 WHITE, H. J.: 'Industrial Electrostatic Precipitation' (Addison Wesley, Pergamon Press, USA, 1963)

9 DARBY, K., and HEINRICH, D. O.: 'Conditioning of Boiler Flue Gases for Improving Efficiency of Electrofilters.', *Staub Reinhaltung der Luft* (in English), 1996, **26**, No. 11, pp. 12–17

10 BICKELHAUPT, R. E.: 'An Interpretation of the Deteriorative Performance of Hot Side Precipitators', *JAPCA*, 1980, **30**, p. 812

11 DURHAM, M. D., *et al.*: 'Identification of low resistivity reentrainment in ESPs operating in dry scrubbing applications'. 9th EPRI/EPA Particulate Control Symposium, Williamsburg, USA, 1991, Session 5A EPRI 100471 2; Palo Alto, Ca., USA, 1992

12 'Chemithon Introduces New Improved Flue Gas Conditioning Systems', *Atmospheric Pollution and Abatement News*, 1993, **1**, No. 1 February, pp. 1–2

13 FERRIGAN, J. J. A: 'Novel Approach to Flue Gas Conditioning'. Proceedings of the EPRI/EPA International Conference on *Managing Hazardous and Particulate Pollutants*, Toronto, Canada, 1995

14 BIBBO, P. P.: 'EPICRON: Agentless Flue Gas Conditioning for Electrostatic Precipitators'. Proceedings of Power-Gen America 94, Orlando, Fl., USA

15 WORACEK, D., *et al.*: 'Catalytic Conversion of Native Sulphur Trioxide in Flue Gas for Resistivity Conditioning'. 10th EPRI Particulate Control Symposium and 5th ICESP Conference, Washington, DC, USA, 1993, pp. 13.1–16; EPRI TR-103048, **2**, Palo Alto, Ca., USA

16 CHANDRAN, R., *et al.*: 'A Review of Flue Gas Conditioning Technology in Meeting Particulate Emission Standards'. Proceedings Electric Power 99, Baltimore, USA, 1999

17 PARKER, K. R.: 'Technological Advances in High Efficiency Particulate Collection'. Proceedings Institute Mechanical Engineers, *Jnl. Power and Energy*, Part A, 1997, **211**

18 FUJISHIMA, H., *et al.*: 'Trend of Electrostatic Precipitator Design in Japan'. Proceedings of the 8th ICESP Conference, Birmingham, Al., USA, 2001, **II**, Paper B 4-1

Chapter 10

'On-line' monitoring, fault finding and identification

In terms of performance, the modern electrostatic precipitator can be designed to operate and produce particulate collection efficiencies in excess of 99.8 per cent. To meet this level of performance, however, it is essential that all elements comprising the precipitator installation operate satisfactorily. As the unit is composed of a large number of mechanical and electrical items, the performance can suffer from various limitations, either because of component failure or changes in the inlet conditions presented to the precipitator. These limitations can be often identified by careful examination and interpretation of the output of installed monitoring equipment and the electrical operating conditions.

10.1 Corrosion condition monitoring

Corrosion in the back end of any plant can be insidious and is usually only found during 'off-line' internal inspection of the plant. When discovered, remedial activities can be very expensive, as not only is the owner faced with replacing or repairing the damaged components, but also the loss of revenue during the necessary plant outage can be significant. For example, assuming a unit charge loss of 2 p kWh^{-1}, then for a 500 MW unit the total cost equates to some £240 000 per day. While a comprehensive consideration of the mechanism of corrosion (normally acid condensation or 'dew point' corrosion attack) is outside the scope of this work, if corrosion occurs either within the precipitator, or in the downstream areas of the plant, its effect can be disastrous, not only in terms of precipitator performance, but also on overall plant efficiency.

Acidic corrosion results from an electrochemical reaction between free acid radicals, moisture and the base metal, with the main effect occurring around the dew point temperature. In boiler plants firing high sulphur content fuels the problems inherent in allowing metal temperatures to fall below the acid dew point temperature are well known, and it is essential to maintain back end temperatures some 20 °C above the theoretical dew point, if acid condensation attack is to be avoided. It has been shown that this 'acid dew point' temperature is dependent on both the sulphuric acid concentration and the moisture concentration in the flue gas as indicated in the following equation [1]:

$$1/T_{dp} = 2.276 \times 10^{-3} - 2.943 \times 10^{-5} \, Log_n P_{H_2O}$$
$$- 8.58 \times 10^{-5} \, (Log_n \, P_{H_2SO_4}) + 6.2 \times 10^{-5} (Log \, _n P_{H_2O})(Log_n \, P_{H_2SO_4}).$$

The effect of SO_3 concentration on the acid dew point temperature is indicated in Figure 10.1 [2], from which it should be noted that for even extremely low SO_3 concentrations the acid dew-point temperatures can be below 100 °C.

If correctly installed and sensibly operated, the majority of ESP installations would be expected to remain reasonably free of corrosion damage. However, if external thermal insulation material is allowed to deteriorate, door seals and/or hopper discharge valves are not maintained, or where the particulate removal efficiency of the precipitator is dependent on flue gas conditioning by injection of sulphur trioxide, chemical reagents or moisture addition, then the plant can be subject to severe corrosion, the rate of which is dependent upon the local gas

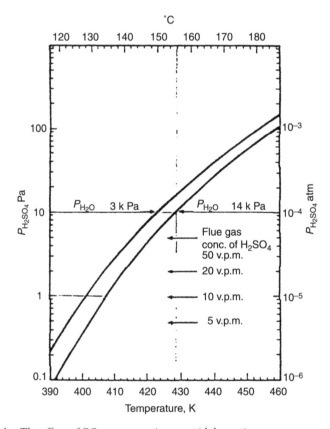

Figure 10.1   The effect of SO_3 concentration on acid dew point temperature

and deposit compositions and substrate operating temperature. In Japan, plants where the recent initiative to mitigate reverse ionisation effects from low sulphur fuels by lowering the precipitator operating temperature to well below a 100 °C [3], could be prime candidates for active dew point corrosion conditions to be established. One unexpected incident was in precipitators following spray dry scrubbers, where in excess of 90 per cent of the SO_2 was removed from the gases but nevertheless severe corrosion was experienced within the precipitator fields.

Although weathered power station fly ashes are quite strongly alkaline, freshly extracted inlet fly ash from sulphur trioxide containing power plant flue gases, e.g. those firing high sulphur coals, can be strongly acidic. Investigations on a plant employing SO_3 flue gas conditioning showed that a 10 per cent w/w suspension of the conditioned fly ash in water was quite acidic, having an initial pH of approximately 3.5, although over a period of some 2 h the pH gradually increased to pH 13. Dust samples from the precipitator outlet as a 10 per cent w/w suspension initially exhibited a pH of approximately 2, rising over some 30 min to a pH of 12. Early investigations attributed the difference in pH effects to the finer particle size and hence higher reactivity of the outlet dust.

Although after a precipitator there may be no measurable free sulphur trioxide in the outlet flue gases, it has been shown that the relatively concentrated (70 per cent) sulphuric acid mist generated earlier in the plant forms complex addition salts with certain compounds, notably ferrous sulphate under lower temperature conditions [4]. Subsequently, the acid absorbs moisture from the flue gas, dissolving its host crystals, and then ionises, thereby promoting corrosion on the carbon steel precipitator components, etc. (This effect is accelerated if fresh fly ash is immersed in moisture, thereby releasing the acid which is subsequently able to react with the alkaline components of the ash, which explains the apparently anomalous changes observed in the pH determinations).

Instead of waiting to carry out an internal inspection of the plant to check for evidence of corrosion, the installation of a modern 'on-line' monitoring system can immediately indicate to the operators the onset of corrosion and hence provide an early warning for the need to change operating conditions to minimise the attack. Because the mechanism of corrosion is electrochemical in nature, a suitable detector located in the suspected (or known) problem areas within the precipitator installation can be used to monitor the plant under all conditions of operation.

To demonstrate the effectiveness of 'on line' monitoring, the plant used in the following example was a 352 MW pulverised coal fired boiler, which could also be fired with natural gas, dependent on the relative costs of the two fuels. The electrostatic precipitators, following Niro type spray driers that readily achieve SO_2 removal efficiencies well in excess of 90 per cent, have experienced extensive corrosion damage ever since they were brought into service. The corrosion was discovered soon after the plant was commissioned and eventually the inlet fields of the precipitators on all three flows had sustained sufficient damage that they had to be de-energised, which resulted in increased particulate emissions. At this stage, it was assumed that the corrosion was associated with poor plant start up and/or commissioning procedures; however, during subsequent annual shut-downs, it was discovered that the second fields were also sustaining corrosion damage similar to that found on the inlet fields. This meant that the problem was associated with continuing plant operation and was not solely the result of commissioning when operating conditions may not have been ideal.

It was suspected that the corrosion damage to the precipitators resulted from short term transient excursions in the spray chamber outlet temperature that allowed moist fly ash and desulphurisation particulates to be carried over into the precipitator. Such moist deposits would agglomerate on the internals during these low temperature periods and could accumulate acidic radicals from the spray dryer outlet, thereby extending the time during which active corrosion could take place. In addition, when such deposits dry out, the concentration of corrosive species would be expected to increase and may cause high rates of acid dew point corrosion to develop within the ESP.

A corrosion monitoring capability was installed within the electrostatic pre-cipitator in order to provide real-time data to establish when corrosion activity was taking place. This required that the sensor system should be sensitive to corrosion and that it should provide the information in a safe and effective manner when the precipitator was energised.

A low temperature electrochemical sensor was designed comprising of six nominally identical insulated electrodes housed in a low flat assembly as shown in Figure 10.2. Electrical connections were made to each element, such that measurements of electrochemical potential and current noise generated spon-taneously from corrosion activity could be collected from three of the electrodes. Polarisation resistance and conductivity measurements were made from a further two electrodes. The two groups of electrodes were separated by an unpolarised 'grounded' electrode, designed to minimise 'cross talk' (i.e. interference) between the polarised and unpolarised monitoring sensor groups.

The detector electrode assembly was installed on a precipitator plate, situated towards the outlet end of the inlet field, approximately one third of the way across the gas flow and 1 m up from the bottom of the plate, as indicated in Figure 10.3 [5]. In this location the detector would attain the same temperature as the plate and would collect pre-charged particulates, so that deposit composition would be closely representative of deposits elsewhere on the precipitator plates. For safety reasons, a nearby section of emitter was removed to minimise the risk of 'flashover' to the detector, and all wires and thermo-couple connections from the sensor were run down the edge of the host plate through a copper conduit line. Considerable care was taken to ensure that this conduit was positively earthed and that conductors leading from the detector to the instrumentation were adequately electrically isolated to ensure protection of personnel and equipment. For example, spark-gap discharge protection and zenner diodes were placed between the precipitator and the monitoring instrumentation.

Figure 10.2   Electrochemical sensor installed on collector plate

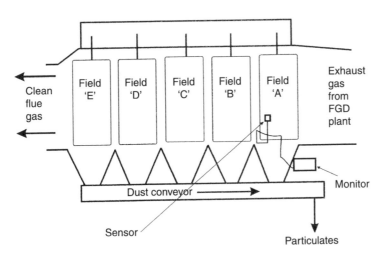

Figure 10.3   Location of detector within precipitator

The sensor probe outputs were interrogated by means of a computerised monitor located close to the precipitator. This monitor comprised digital modules for logging the corrosion activity and temperature response of the sensor in order to establish under what operating conditions the precipitator was at risk. Measurements on the individual modules were made sequentially, at intervals of 10 min, to avoid interference between polarised and unpolarised electrodes on the detector.

Figure 10.4 is a summary graph showing the overall changes in corrosion activity during the period from September 1997 to January 1998. The signals are somewhat noisier than might be expected; this is undoubtedly because the data is highly compressed and contains gross short term variations associated with changes in plant conditions. From this, two important trends become apparent. Initially there was a tendency from September to December 1997 for corrosion activity to increase as the probe 'conditioned' (i.e. as it accumulated acid and corroded to produce deposits that would increase the surface conductivity when moistened with acid or condensation products); and then, during January 1998, there was a significant decline in recorded corrosion activity. With the decrease in activity during January, it was suspected, and later confirmed, that the earlier increase in corrosion activity had been recognised by operational personnel and measures had been introduced to counteract it.

Figure 10.5 illustrates the activity pattern that was typical of the transition from gas to coal firing during a plant start up. The high initial gas temperature (upper trace) is typical of the gas firing condition, during which corrosion activity is low. The step fall in temperature at around 7.00 a.m. on the 9th September and the corresponding increase in corrosion rate marked the introduction of coal firing and commissioning of the spray drier plant. Over the course of the

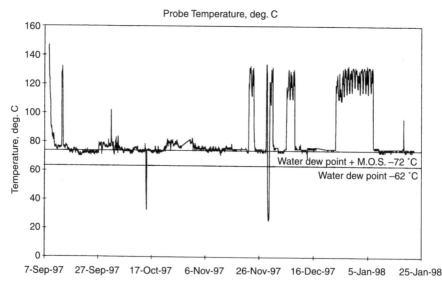

Figure 10.4   Trend in corrosion activity from September 1997 to January 1998

next 18 h there was a gradual increase in corrosion rate activity before the plant stabilised and the corrosion rate subsided towards the background levels.

Figure 10.6 shows the conditions during a transition from coal to gas firing, then a plant shut down, then re-firing on gas, and finally the resumption of coal firing. Initially on the 26th November, the plant was on-line firing coal, but on the 28th November the plant was switched to gas, which is marked by the sudden rise in flue gas temperature, accompanied by a peak in the corrosion rate. The fall in temperature to ~20 °C and the concurrent fall in corrosion rate is

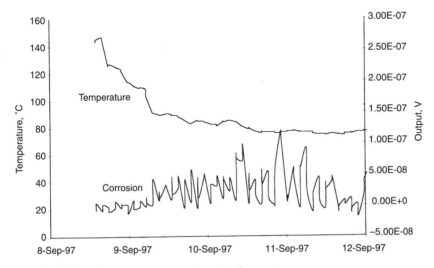

Figure 10.5   Trend in activity pattern during boiler start up

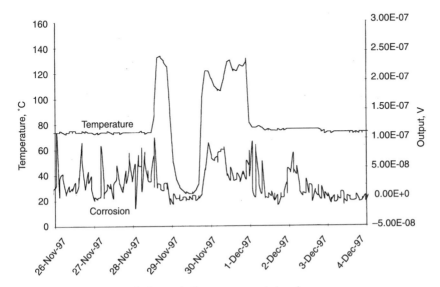

Figure 10.6   Activity trends during boiler start up and shut down

associated with the plant being taken out of service. On the evening of the 29th November, the plant was put back on line under gas firing, as indicated by the increasing gas temperature rising to a peak on 30th November, and an associated increase in corrosion activity. The step reduction in operating temperature at around 8.00 a.m. on the morning of 1st December, together with an associated reduction in corrosion rate, corresponds to the plant switching back to coal firing. Thereafter the corrosion rate fell until it stabilised at background levels during the 3rd and 4th December 1997.

This pattern of start up and shut down activity behaviour is further illustrated in Figure 10.7, where the temperature fluctuations shown by the upper curve are reflected by the (relatively) consistent transients in corrosion rate exhibited by the lower curve. Actual corrosion rates are dependent upon the prevailing plant operating condition and subsequent detailed reference to plant control room data revealed just how closely this was affected by operational constraints.

When the unit was on gas firing, on line corrosion activity was consistently recorded and the rate tended to be higher than that observed immediately beforehand under coal firing. This could be attributable to cleansing of the SCR catalysts after coal firing, the operating procedure being adjusted to continue gas scrubbing for a short time after changing from coal to gas in order to avoid inadvertent sulphur dioxide emissions.

Unexpectedly, there is no doubt that precipitator corrosion is exacerbated by the combination of both natural gas and coal firing. During coal firing, sulphur and chloride products are present, and these remain within the system when gas firing commences. Under gas firing, the precipitator flue gas temperature is higher and contains an increased moisture content. This means that deposits within the precipitator tend to 'sweat' and release absorbed acid; this sweating

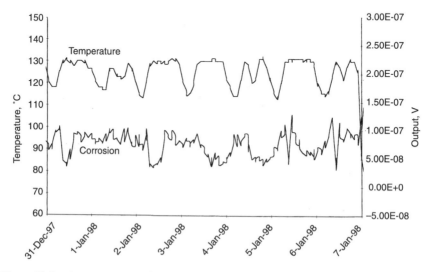

Figure 10.7   Activity pattern during the monitoring investigation

allows the degree of ionisation of the acid to increase rapidly (making it more corrosive), and the carbon steel components of the precipitator are subject to corrosion until all freed acid is neutralised. Furthermore, with gas firing, the spray driers are normally out of service thereby limiting the amount of calcium based particulates available to neutralise the acidic condition. The investigation showed conclusively that alternate coal and gas firing meant that acidic deposits formed during coal firing remained in the plant when it switched to gas and a relatively high proportion of the residual acid remained available to corrode the steelwork. In consequence, the overall rate of attack sustained by a coal and natural gas dual-fuelled plant is much higher than would be expected for either of the fuels if used separately.

From the graphical data it was evident that there were several regimes when corrosion took place and the findings of the monitoring investigation can be summarised as follows:

- during plant warm up owing to low flow/low gas temperatures;
- during spray dryer start up as a result of low spray drier exit temperatures;
- during generation under coal and associated low spray drier outlet temperatures;
- during generation under gas, owing to the presence of acidic deposits from coal firing being reactivated by the higher gas moisture contents.

After the previous history of serious corrosion damage to the precipitators, reported to have been typically several millimetres per annum throughout the previous decade, it was a surprise when the instrumentation revealed that, after a few months of actual monitoring, very little corrosion appeared to be taking place. Although fluctuations in the corrosion rate were evident on the trend graphs, these were clearly related to changes in the operating regime of the boiler. The results indicated the overall rate of attack to be much lower than had been expected, the average annual corrosion rate on carbon steel components in the precipitators being only 5–50 μm.

The apparent reduction in corrosion activity was confirmed at the first annual inspection by the relative lack of corrosion product on the sensor surface and the absence of the agglomerated dust deposits and build up that had been noted prior to installation of the monitoring instrumentation. The minimal rate of corrosion attack was actually verified by measurements of collector plate thickness made independently by plant personnel during the shut down. Subsequent discussions with the station personnel revealed that, based on the monitoring data, the temperature indication obtained from the corrosion sensor was now used to optimise control of the spray drier chamber outlet temperature, rather than the supplier installed device. In effect, this meant that the spray drier outlet temperature was raised slightly during the early period of FGD plant operation. In addition, operations personnel had increased slightly the rate of lime addition to the FGD spray liquor circuit to increase SO_2 removal efficiency. These measures would readily explain the observed reduction in precipitator corrosion damage.

With the satisfactory outcome of the above investigation, the monitoring installation has been modified to provide a full 'on-line' indication of precipitator condition within the main plant control room via the process control computer display. The purpose of the system is to provide an early warning to operations personnel of when the precipitators are at risk of attack, such that minor adjustments to normal operating parameters can be made to mitigate any corrosive environment.

In general the principle of 'on-line' monitoring can prove advantageous to operational plants which may be subject to potential corrosion and thereby prevent emission excursions due to resultant damage to the precipitator internals or downstream equipment.

10.2 Electrical operating conditions

The areas within a precipitator installation where typical difficulties arise are shown in Figure 10.8. In identifying these areas it is assumed that the precipitator installation has been satisfactorily erected and commissioned [6] and that a performance problem, as evidenced by the opacity meter, develops during normal plant operation.

As with any item of equipment, if problems arise, it is important that a systematic approach be adopted to identify and rectify the situation. The areas that are required to be examined are:

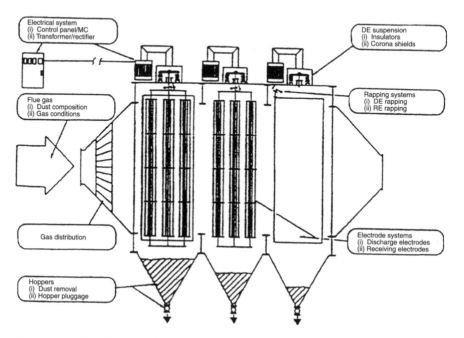

Figure 10.8   Identification of potential precipitator fault areas

(1) rectifier equipment,
(2) deposit removal from the ESP internals,
(3) electrical characteristic curves,
(4) DE and collector alignment,
(5) high voltage insulator systems,
(6) hoppers and dedusting,
(7) gas distribution and air inleakage,
(8) changes in gas inlet conditions.

It is not suggested that these are the only areas where difficulties may arise that impact on the overall performance of the unit, but the following checks can often identify the likely cause of the problem and, more important, most can be made without having the unit off line. Consequently they should be considered as the first approach to any precipitator fault finding procedure, that is, 'on-line' monitoring.

Although problems associated with the transformer rectifier equipment are rare, the signal lamps and meters can give invaluable information as to a likely cause of the difficulty being experienced. While the following analysis applies specifically to conventional single phase mains rectified electrical equipment, many of the symptoms equally apply to other forms of energisation but not necessarily mirrored by some of the examples shown.

10.2.1 TR control cubicle information

Table 10.1 summarises the most common electrical problems together with their symptoms as indicated on the TR control panel itself.

10.2.2 TR set panel meter readings

Although the foregoing identifies difficulties that may prevent the TR set being energised, Table 10.2 assumes that the field has been successfully energised and the panel meters are operating satisfactorily.

Examination of the TR panel meter readings can often indicate the likely fault, which makes location and rectification easier. Again the following specifically refers to precipitators that are energised by conventional mains rectified equipment, although some of the symptoms can equally apply to other forms of energisation.

10.2.3 TR equipment lamp test

Although the instrumentation on the TR equipment can provide specific data on the operational condition of the installed equipment as indicated in the foregoing section, it assumes the equipment is connected to a field. If, however, the equipment is not connected, or even if it is, there is another fairly simple method of checking the equipment using a low voltage supply and a standard light bulb.

Table 10.1 Common electrical problems and their symptoms

Location	Symptom	Likely fault
TR set control cubicle	Green lamp fails to light on closing isolator.	(a) No supply voltage (b) Bulb filament burnt out. (c) Auxiliary contacts on isolator not making. (d) Auxiliary isolator fuse O/S.
TR set control cubicle	HT 'on' contactor fails to close on operating switch.	(a) Contactor failing to close correctly. (b) Contactor fuse O/S. (c) TR set 'tripped', e.g. high oil temperature. (d) Local/remote selector switch in remote condition.
TR set control cubicle	Red lamp fails to light when HT contactor is closed.	(a) Bulb filament burnt out. (b) Auxiliary contacts not 'making'. (c) Auxiliary fuse O/S.
TR set control cubicle	Contactor trips on closing.	(a) No fluid in dash pots. (b) Overload setting too low. (c) Dead short on electrode system.

TR set control cubicle	Contactor trips. High mA but low kV.	(a) Partial or high resistance short of electrode system;
		   – hopper dust level.
		   – HT insulators tracking.
		(b) Dust deposits on electrode system.

TR set control cubicle	Contactor fails to trip on continuous overload short.	(a) Check if dash pot fluid is correct viscosity. (b) Check if dash pot piston release hole is clear.

Rather than fully energising the equipment from a supply of ~400 V a.c., the transformer primary is energised from a standard 110 V a.c. outlet having a 100 W tungsten filament light bulb connected in series with the transformer input terminal. Using a 100 W lamp limits the line current to 1 A even under short circuit conditions on either the transformer or equipment. Although the equipment is only energised from a low voltage source, the secondary output voltage can still be some 20 kV and the test should only be carried out, under strict

Table 10.2 TR panel meter readings and likely faults

Meter readings				Likely fault
Primary volts V	Primary amps A	Secondary volts kV	Secondary amps mA	
Zero	Zero	Zero	Zero	(a) Mains supply fault. (b) Thyristor fault. (c) AVC fault. (d) a.c. inductor O/C if fitted. (e) Control circuit fault.
Zero or low	High	Zero or low	High	Possible short circuit on: (a) precipitator, misalignment, dust short or build up, etc., (b) high dust level in hoppers, (c) earth switches 'closed', (d) HT cable or connection down, (e) internal TR fault – shorting to earth.
High	Low	High	Zero	Possible open circuit on: (a) HT feed lines, (b) HT changeover switches if fitted, (c) internal TR fault – open circuit.
Zero or low	High	Zero or low	Low	(a) TR set fault of some type, e.g. primary short.
Normal	Normal	Zero	Normal	(a) Voltage divider fault. (b) Meter fault. (c) Wiring O/C.
Zero	Normal	Normal	Normal	(a) Meter fault. (b) Fuses O/C on meter if fitted.
Normal	Zero or low	Normal	Normal	(a) Current transformer fault. (b) Meter fault. (c) Wiring fault.
High or normal	Low or normal	High	Low	(a) Dust deposits on discharge electrode system. Check rapping.
Normal	Normal	Normal	Zero	(a) Faulty meter. (b) Shunt short circuit.
Indication of excessive arcing on either manual or automatic control.				(a) Precipitator fault, e.g. broken element.

Table 10.2 continued

Meter Readings				Likely fault
Primary volts V	Primary amps A	Secondary volts kV	Secondary amps mA	
				(c) Cracked/contaminated rapping drive insulator.
				(d) Transformer fault condition.
				(e) Change in inlet gas conditions.
A higher kV and or performance can be obtained under manual rather than automatic control.				(a) The AVC requires setting up to meet operating gas conditions or is faulty.

safety procedures, by persons familiar with high voltage equipment testing, after ensuring that the output terminal is 'electrically safe'.

Examination of the light intensity of the bulb can be used to indicate the following conditions.

(a) Very dull glow indicates:

    (i)   the equipment is 'healthy' in that there are no shorted turns on the transformer,

    (ii) if the outlet terminal is connected to a field, the field does no exhibit a short,

    (iii) the rectifier bridge is satisfactory, i.e. not shorted.

(b) Bright glow indicates:

    (i)   the transformer has shorted turns on the primary,

    (ii) the rectifier diode bridge has a short,

    (iii) if connected to a field, then there is a short on the field or feed connection.

(c) No glow indicates:

    (i)   open circuit in the primary connections of the transformer,

    (ii) if a linear reactor is connected, a fault in the connections,

    (iii) there is no voltage supply to the system.

This lamp test, although useful as a simple means of checking the condition of a rectifier set, should, if a short or open circuit situation is indicated, be followed with a fuller investigation to determine the fault location.

10.3 Deposit removal from the internals

For the precipitation process to proceed unhindered, it is important that the internal components are kept relatively free of deposits. On a dry precipitator this is usually achieved by mechanical impact rapping of some type, whereas, for a wet installation, liquid washing is adopted to maintain the internals in a clean condition.

Unless the system is successfully operating and removing deposits both from the collectors and discharge electrode system, the deposits can adversely impact on the electrical operation and hence reduce performance. (The actual mechanism of how the deposits impact on the electrical operation and hence performance was discussed in previous chapters.)

In general, the problems that can arise impacting on performance are as follows. Deposits on the collector plates, which receive the majority of the material, reduces the electrical clearance and hence flashover voltage and therefore performance. On a dry precipitator installation, depending on the electrical resistivity of the particulates, deposits can produce a significant voltage drop, which impacts on the electric field distribution and particle charging. Deposits on the discharge elements can significantly reduce the corona current generation, because the radius of curvature of the element is increased, which can affect particle charging and hence performance.

If there is a significant fall in electrical operating conditions from the 'clean plant' V/I characteristics (see Figures 10.9 and 10.11), then examination of the rapping control panel instrumentation can often indicate the likely fault, which makes subsequent location and rectification easier; see Table 10.3. Where the rapping is computer controlled, motor indicator lights can be used to check operation.

Although Table 10.3 refers to rapping on a 'dry type' precipitator, in the case of a 'wet type' precipitator installation there will be an equivalent water flow control panel for cleaning the internals. This will have pressure and flow control switches, etc., but the interpretation of likely faults will take a similar format.

A difficulty that may arise on a spray irrigated type of wet installation occurs if one or more irrigation sprays become blocked and jets rather than atomised water are ejected. Should the jet impinge onto the discharge electrode system, then a short circuit situation can readily develop producing a fall in performance. In this situation switching off the water supply can often identify this problem.

10.4 DE and collector system voltage/current relationships

As a diagnostic tool, the V/I curve can provide invaluable information as to possible faults within the precipitation field. There are three types of curve to be considered: the first is under static air conditions, the second with process gas

Table 10.3 Rapping control cubicle

Location	Symptom	Likely cause
Rapping motor starter panel	Motor fails to start	(a) Check selector switch position. (b) Check if 'stop' buttons locked off. (c) Check electronic shear pin setting if fitted.
Rapping motor starter panel	Motor 'trips' immediately	(a) Check rapping shafts not 'jammed'. (b) Check gearbox for jamming. (c) Check overload settings on electronic shear pin, if fitted.
Rapping control panel	Motors do not operate in correct timing sequence	(a) Check PLC or computer settings.
Rapping control panel	Discharge electrode rappers not operating	(a) Check for broken drive insulator. (b) Check gearbox.

passing through the plant and the third under dirty conditions without process gas, but with hot gas conditions prevailing.

10.4.1 Clean air load characteristic

The first relationship is normally termed an 'air load test'; and the initial test should be carried out during plant commissioning. The resultant information forms the basis on which later comparative assessments can be made. To make a V/I curve, the discharge currents are recorded for, say, 5 kV increments of secondary voltage; at the same time it is useful to also record the corresponding primary voltages and currents for each incremental step for later reference.

The corresponding current for each incremental step is then plotted until the flashover voltage occurs, or the secondary current or voltage limit is reached. With air, the low space charge that prevails usually means that it is the current that is the limiting condition, rather than secondary voltage. If possible, when carrying out this air test the access doors should be left open, to avoid the build up of ions within the precipitator, which can impact on the final measured values. The resultant curve, as shown in Figure 10.9, provides the basic electrical foot print of the precipitator field. If all fields have the same type of discharge element and collector spacing then the curves for each field or precipitation zone should be similar in terms of corona inception and breakdown voltage. If the curves indicate serious deviations, then it usually means that the alignment is faulty or there is an internal or external electrical clearance between the

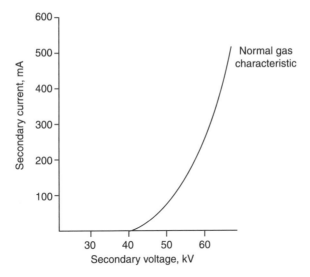

Figure 10.9   Static V/I characteristic of precipitator field

energised sections and earth, which adversely impacts on the operating conditions.

10.4.2 Operational curves with gas passing through the system

These operational curves under process gas conditions are determined in a similar manner to the basic air load test by slowly raising the operating voltage and plotting the corresponding secondary current flow until flashover or the set limits are reached.

In the case of operational characteristics with dusty gas passing through the electrode system, the effect of space charge becomes evident; as the gas passes from the inlet to the outlet of the precipitator, the particle concentration decreases which reduces the space charge effects and the corona current flow significantly increases. This effect is illustrated in Figure 10.10, for a conventional three field precipitator handling a 'normal' precipitable dust, that is, one having a particulate electrical resistivity in the order of 10^{10}–10^{11} Ωcm, where the increase in current in the outlet field is clearly shown. As the electrical breakdown voltage will be similar for the same collector spacing and electrode system, the mean voltage as shown on the kV meter will be lower than upstream fields, owing to the increasing voltage ripple in supplying the increased current.

The characteristic curves shown in Figure 10.11 illustrate some shortcomings and operating restrictions arising because of mechanical limitations on the precipitator. In all cases, because the electrical readings are compromised the collection performance will be reduced.

Curve A_1 illustrates a resistive path to earth being a straight line. This may be the result of a contaminated HT support or rapping drive insulator or a high

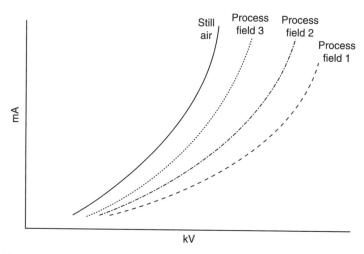

Figure 10.10   Typical V/I characteristics for 'good' fly ash

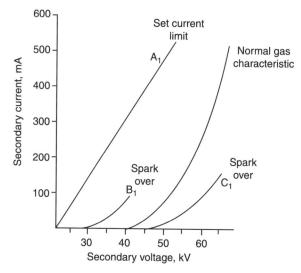

Figure 10.11   V/I characteristics indicating operational performance problems

hopper dust level. (If the resistance path is low it may not be possible to energise the TR set at all.)

Curve B_1 illustrates severe mal-alignment between the collectors and discharge electrodes, being characterised by a low corona onset voltage followed by breakdown at a low secondary voltage. This type of curve can also be the result of excessive collector dust build up.

Curve C_1 represents the type of characteristic attributable to heavily built up discharge electrodes, which is raising the corona onset voltage through increasing the effective diameter of the discharge elements. In this event it is normal for the breakdown voltage to be significantly reduced, depending on the degree of build up experienced.

Abnormal characteristics are illustrated in Figure 10.12, and these usually arise because of having to handle a 'difficult' dust, that is, one having an electrical resistivity greater than $10^{11}–10^{12}$ Ωcm.

Curve A_2 represents the type of characteristic arising with severe reverse ionisation. This occurs when the resistivity is so high that the particle negative charge is unable to leak to earth, resulting in a voltage build up on the surface of the deposit, which gives rise to the production of positive ions. These leave the collector and migrate towards the discharge electrodes, adversely modifying the electric field pattern and neutralising the negative corona flow. Under this condition emissions have been known to increase by at least a factor of 10.

Curve B_2 indicates the characteristic of a high resistivity dust, but without full reverse ionisation developing; in this case, as with severe reverse ionisation, the TR set current limit may be reached before electrical breakdown occurs.

Curve C_2 represents the electrical operating characteristics for a dust having a borderline electrical resistivity, where instead of positive ions developing and leaving the deposit, the voltage across the deposit increases only to a point where breakdown occurs, the value of voltage being dependent on the resistivity of the material.

In examining the operational curves it is sometimes difficult to distinguish between the characteristic arising because of electrode mal-alignment Curve B_1, that due to deposit build up (Curve C_1), and that resulting from a moderately high electrical resistivity fly ash (Curve C_2). Comparing these curves with those taken immediately following commissioning, when alignment is good and there

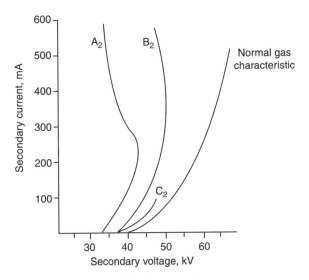

Figure 10.12   Abnormal operational V/I characteristics

is no serious deposit build up, particularly when handling a low resitivity fly ash, should indicate which symptom is the most likely.

10.4.3 Dirty air load test, without gas passing through system

Occasionally it may not be possible to inspect the unit internally, even without gas passing through it. In this case an air load test taken under hot conditions can, nevertheless, provide useful information as to the likely cause of changing electrical characteristics, in addition to the increase of the corona current resulting from the increased temperature.

Figure 10.13 shows the impact on the electrical characteristics of dust deposition on a dirty plate compared to a clean plate under ambient conditions. Generally the operational voltage with the clean collector (A_3) appears higher than for a dirty collector plate (B_3) for similar corona conditions, because of the voltage drop across the deposited layer. When comparing the inlet, centre and outlet field characteristics, the corresponding curves under dirty conditions may not be identical, even if the alignment is acceptable, because the resistivity of the deposit and hence voltage drop can vary with particle size, dust composition and humidity of the air.

Compared with the process gas characteristics represented in Section 10.2, these dirty air tests tend to lie on the left side of the axes, with higher currents, mainly as a result of the reduced space charge effects and a higher voltage level because of the drop across the deposited dust layer.

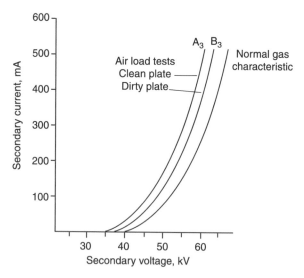

Figure 10.13   Comparison of V/I characteristics for clean and dirty collector plates

10.5 Collector/discharge electrode alignment

Figure 10.11 indicates in curve B_1 that the electrical characteristics can be pre-determined by the alignment of the internal components. Generally with the height of collectors and discharge electrodes in common usage for large precipitator installations, the optimum alignment that one can achieve in practice, because of fabrication and erection tolerances, is around 10 mm. For plants having collector spacings of 400 mm, this means that the alignment can be out by around 5 per cent which reduces the optimum operational voltage by a similar amount; for closer collector spacing the optimum voltages are reduced even further. It should be remembered that the flashover voltage is determined by the lowest electrical clearance so even one discharge element being mis-aligned has a significant effect on the operating voltage and hence performance.

Although discharge electrodes in current usage are designed to have a long life, occasionally, one may break or come adrift from the frame. If this occurs, then the free end is drawn towards the collector plate by the electric field. As it nears the plate flashover occurs and the electrode tends to fall back under gravity. Once the field is re-established the free end is again drawn towards the collector and flashover recurs. This collapsing and re-establishing of the field following breakdown results in a cyclical pattern of power on, power off, the periodicity of which is determined by the free length of the element that is swinging.

One of the reasons for carrying out an air load test following commissioning, or following internal work on the unit, is to ensure that not only are the electrode alignment and internal clearances satisfactory, but that nothing has been left inside the unit that could later impact on the electrical operation. Items that have been found are welding wires used to support lamps, lengths of rope, etc., which tend to lie in a vertical plane but, when energised, stand out horizontally across the inter electrode region, significantly reducing the flashover voltage.

10.6 High tension insulators

Operational difficulties can arise with the HT support or lead through insulators used for isolating the discharge electrode system from the casing, which is at earth potential. Generally these difficulties arise through contamination of the insulator surface with flue gas components, typically, dust, moisture or sulphuric acid deposition. If deposition occurs then, dependent on the degree of contamination and its resistivity, the secondary voltage can short out through the deposit, initially leading to a very low operating voltage and high current discharge, which, depending on the short circuit current, can shatter the insulator.

To avoid these it is normal practice to purge the insulator with hot air, such that the insulator, whether it is for discharge electrode support or rapping, remains out of the gas stream and the surface ideally remains above any potential acid dew point temperature. If the heating or purging system fails, then the insulators are subject to possible contamination and can produce a characteristic curve similar to that shown as curve A_1 in Figure 10.11.

10.7 Hoppers

In any operational unit, build up in the collecting hoppers can cause performance short falls, usually because of limitations resulting from the discharge or evacuation systems failing to remove the collected material. If this occurs then the dust level can build up and will eventually short out the discharge electrode system. Depending on the build up resistance, the form of characteristic curve is as depicted as curve A_1 in Figure 10.11.

Unless the build up situation is speedily addressed and dust builds up between the collector plates to reach a height of about 1 m, and a 'rat hole' develops during evacuation (that is, the dust forms a hole rather than falling uniformly through the hopper), because freshly collected dust is relatively fluid, a large hydrostatic pressure can develop that exerts sufficient load on adjacent collectors to distort them. This can be such that, even after reinstatement of the evacuation system and emptying of the hoppers, the damage and mis-alignment to the internals can significantly impact on operational voltages and hence performance.

An operational situation to be considered is that, if the inlet field evacuation system fails, then de-energising the inlet will not necessarily prevent further hopper deposition, because the coarser and heavier particles reaching the precipitator will have been decelerated sufficiently to fall out under gravity. Unless the hopper is cleared then the dust level can reach a height to damage the internals as indicated above

Although the largest quantity of dust will be deposited in the inlet field hoppers, this will also have the coarsest particle sizing and hence lowest cohesive interstitial forces and, therefore, be the easiest to discharge. Dust collected further down stream will have progressively smaller and narrower particle sizings and will be more difficult to evacuate in spite of the smaller mass. This needs to be carefully addressed in the design and during process commissioning to ensure evacuation of all hoppers is satisfactory.

To prevent untreated hopper gas bypass impacting on the performance, steps are taken in the gas distribution system to prevent this occurring; as a result, the gas and material in the hopper is heated only by radiation from the horizontal gas stream. As the hoppers have relatively large surface areas, to prevent the walls cooling off, possibly to below a dew point temperature situation, heaters are generally applied to match the heat loss through the thermal insulation. Keeping the dust hot and dry is essential to minimise dust pluggage, bridging and other hopper dedusting difficulties. On some plants air distribution pads, which inject warm air into the hoppers, are used to fluidise the dust to assist its discharge. Although vibrators have been fitted to some installations, these can result in compaction of the dust and actually make the dedusting situation worse.

Because of the serious performance short falls associated with hopper dedusting problems, it is imperative that during any investigation the hopper dedusting system and hopper dust levels are checked out and modified as necessary. Routine checks on the installed instrumentation and monitoring systems should assist in minimising precipitator performance difficulties.

10.8 Gas distribution and air inleakage

Because there is an inverse exponential relationship between efficiency and gas volume, hence gas velocity, it is important, as discussed in Chapter 3, for the gas distribution to be maintained at the correct standard. Although the gas distribution through the unit should have been checked and confirmed prior to full scale commissioning, it is possible during operation for the inlet ducting, in particular, to become partially blocked by dust deposition, which can upset subsequent gas distribution patterns. Although it is necessary for the plant to be off line and cold, an internal inspection will reveal if duct deposition is having an adverse effect of gas distribution and hence performance.

Electrically, although it is difficult to assess if duct build up is causing a gas distribution problem, if a hole in the casing develops or an access door inleaks cold air, typically as a jet several metres in length, a local gas distribution upset is invariably produced. This cold air inleakage usually results in rapid electrical flashover at a voltage level much lower than would be anticipated from the characteristic curves as a result of the change in gas composition and local gas temperature.

Air inleakage into the hopper region, either through ill sealing doors or leaking joints in the evacuation system, can not only re-entrain previously collected and deposited dust, but in doing so can entrain this dust directly into the outlet duct at a high dust concentration, which can seriously impact on the emission and hence performance. An air inleakage into the outlet field of <1 per cent of the total flow can result in the emission doubling as a result of entrained hopper dust.

10.9 Changing inlet gas conditions

10.9.1 Particle resistivity

The most significant change in electrical operating conditions typically arises with changes in the electrical resistivity of the particulates presented to the precipitator. With boiler plant applications handling fly ash from different coals, the electrical resistivity is largely determined by the sulphur content of the coal plus the sodium oxide content of the ash, see Chapter 9. As the electrical resistivity increases, the electrical characteristics alter as indicated in Figure 10.12, curves A_2, B_2 and C_2. By modifying the resistivity by the injection of chemical

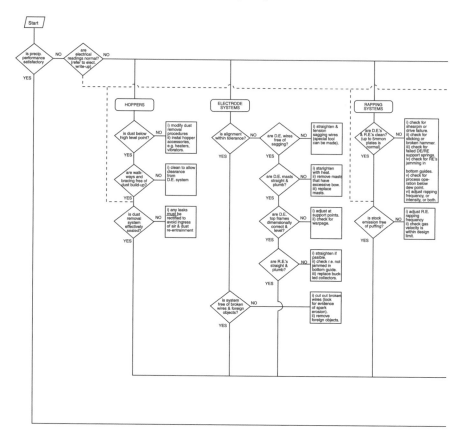

Figure 10.14   Systematic fault finding procedure

conditioning agents, it is possible to change the electrical operating conditions back towards a more normal characteristic.

10.9.2 Particle sizing and opacity monitoring

Examination of the opacity chart can also give invaluable information as to the likely cause of performance degradation. Although reductions in the electrical operating characteristics, as reviewed above, will result in decreases in performance, if the particle size changes with different fuels or conditions, then the opacity traces can indicate the following.

(a)   The emission increase may be a result of a general shift to a lower outlet particle size range, which may not alter the collection efficiency because many opacity meters are particle size sensitive. But care should be exercised

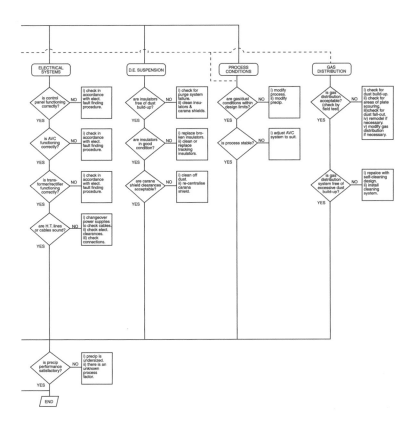

in reaching this conclusion, because dependent on the system of measurement the variation may not be the result of a reduced mean particle size. For example, a light transmission type of device is particle size sensitive because the absorption coefficient is dependent on the number of particles present in the light beam, whereas for back scatter devices the surface area of the particles is more important than their size.

(b) A cyclic 'spiky' trace can be the result of particle re-entrainment through the particulates having poorer cohesive properties and consequently on rapping, are released to pass into the outlet ductwork and as such have a serious impact on the overall collection efficiency. In this situation one must carry out a full rapping investigation by altering the rapping cycle systematically and observing the resultant opacity traces until a satisfactory trace is obtained.

10.10 Systematic fault finding procedure

Although the foregoing sections generally relate to the identification of possible faults using the electrical operating conditions and other plant instrumentation, they must be considered only as the initial step in any investigational procedure. Although some likely faults will be readily identified using the above approaches, it will be eventually necessary to have the unit off line to carry out an internal inspection and to perform corrective work on some of the more serious faults in order that optimum precipitator performance can be re-established.

Figure 10.14 represents a systematic precipitator fault finding schedule that can be used for most installations, remembering that for wet precipitators, the cleaning/deposit removal systems are liquor based.

The main purpose of this chapter has been to indicate how an examination of the installed monitoring equipment and instrumentation can be effective not only in identifying possible faults but also in optimising the performance of the precipitators. With the introduction of more powerful supervisory computer control systems, as reviewed in Chapter 6, being installed as a norm, many of the electrical and other operational difficulties discussed above can now be identified from in-built fault finding menus.

10.11 References

1 BANCHERO, J. T., and VERHOFF, F. H.: 'Evaluation and interpretation of the vapour pressure data for sulphuric acid aqueous solutions with application to flue gas dewpoints,' *Jnl. Inst. Fuel*, 1975, p. 76

2 HALSTEAD, W. D., and TALBOT, J. R.: 'The Sulphuric Acid Dewpoint in Power Station Flue Gases', *Journal of Inst. Energy*, 1980, pp. 142–6

3 FUJISHIMA, H., *et al.*: 'Colder Side Electrostatic Precipitator of Advanced Flue Gas Treatment System for Coal-Fired Boiler'. Proceedings of the 7th ICESP Conference, Kyongju, Korea, 1998

4 Holmes, D. R. (Ed.): 'Dew Point Corrosion' (Ellis Harwood Ltd, 1985)

5 COX, W. M., and PARKER, K. R.: 'Performance and Condition Management at Durnrohr Power Station'. Paper Presented at the 1998 European User Group Conference, Lea Hall, Warwickshire, UK, June 1998

6 Parker, K. R. (Ed.): 'Applied Electrostatic Precipitation' (Chapman & Hall, London, 1997, chapter 11)

Author index

Anderssen C. 19
Ashton M.D. 62

Banchero J.T. 264
Baylis A.P. 36, 87
Bickelhaupt R.E. 237
Bibbo P.P. 237

Casadei D. 214
Caputo A.C. 172
Chandran R. 237
Cho J.G. 214
Cochet R. 27–8, 36, 46, 61
Cohen L. 236
Cottingham C.R. 86
Cottrell F.G. 68, 86, 92, 117, 172
Coors Insulator Pub 87
Cox W.M. 264
Cristescu D. 173

Dalmon J. 47, 59–62
Darby K. 87, 236–7
Deutsch W. 33–4, 37, 46, 61
Devine P. 213–14
Dickenson R. 236
Dubard J.L. 172
Durham M.D. 237

Elholm P. 173
Elsner R.W. 62

Falaki H.R. 87
Farley R. 59, 62
Feldman P.C. 171–2
Ferrigan J.J. 237
Fluent Inc (CFD Prog). 9, 53
Fujishima H. 172, 237, 264

Gaugain J.M. 22, 36
Gibson D. 62
Goard P.R.C. 230
Grieco G.J. 19, 86
Guitard C.F. 89, 117

Hall H.J. 172, 237
Halstead W.D. 264
Hewitt G.W. 27, 36
Hein A.G. 62
Heinrich D.O. 237
Herder H. 214
Holfeld M. 89, 117
Holmes D.R. (Ed) 264
Houlgreave J.A. 87

ICAC Publication EP7 19, 62
IEEE Publication (Std 548)
    236

Jacobsson H. 146

Kauppinen E.I. 62
Klinger E.H. 59, 62
Kirsten M. 214
Kumar S. 62

Lind L. 19
Liu B.Y.H. 36
Lodge O.J. 21, 36, 68, 89, 91,
    117
Lowe H.J. 47, 59, 61–2, 87
Lausen P. 172
Lucas D.H. 59, 62, 87
Lund C.R. 214

Matts S. 34, 37
Mauritsen C. 172
McDonald J.R. 46, 61
Mohan N. 214
Murphy A.T. 36

Nakaoka M. 213
Neulinger F.B. 171
Nichols G. 236
Nielsen N.F. 62
Novagoratz. D.M. 62

Ohnfeldt P.O. 34

Parker K.R. 19, 62, 86, 237,
  264
Penney G.W. 59, 62
Perez M.A. 214
Petersen H.H. 37, 172
Petkov R. 214
Porle K. 146, 171–2, 214
Potter E.C. 236

Ranstad P. 214
Reyes V. 146, 214
Roentgen V. 22, 36
Russell-Jones A. 87, 171

Sanyal A. 62
Schmoch M. 172
Schioeth M. 172
Schwab M.J. 19
Seitz D. 214
Siemens web site 213
Smith W.B. 36, 61
Spencer H. 236
Srinavasachar S. 214
Star C.D. (CFD Prog) 52, 62

Talbot J.R. 264
Tali-ighil 214
Terai H. 172
Thomsen H.P. 46, 61
Tidy D. 59–60, 62
Townsend J.S. 22, 24, 36
Trichel G.W. 36

UK EPSRC Grant 214

Valentin F.H.H. 59, 62
Verhoff F.H. 264

Walker A.O. 117
Weinmann S.F. 146
Wernekinck E.A. 213
Wetgen D. 19, 171
Wetteland O. 214
White H.J. 87, 172, 237
Wizerani M. 213
Woracek D. 237

Yamamura A. 173
Yoshikazu N. 173

Subject index

Page numbers appearing in **bold** refer to figures and page numbers appearing in *italic* refer to tables.

Aerodynamic factors affecting performance
    44–51
    See also gas distribution.
Aerosols 57, **57**, 58, 215
Agglomeration 5, 32, 54–6, 60
Air inleakage hoppers & ducts 261
Applications of ESPs. 2, 10–11
Application Range of ESPs 1–2
Automatic voltage control
    analogue types – Prim. Current 113, 131
    Sec. Voltage 114, 130
    microprocessor – A/D inputs 115, 129
    Early closed loop designs. 136
    back corona detection 140, **141**
    supervisory type gateway 143, **144**
    advanced function AVCs 144, *145*

Back corona, *see also* Reverse Ionisation
    onset of back ionisation 15
    severe back ionisation **83**, 140
Brownian motion 27, 31
Bus section/field 5, 75–6

Carbon absorption of heavy metals etc. 58
Carbon particles 14, 16, 48, 58, **60**, 216–17,
    226
Casings, steel, brick, GRP etc. 41
Cenospheres 18, 216
Charles Law 41
Cochet charging model 27–8
Cohesion/cohesivity 56, 58, 60, 226, 235
Cold side ESPs 220, 224
Collector elements
    catch space 72, **72**
    concentric ring 8
    GRP 8, 73
    hexagonal 8
    rolled channel 72, **72**
    tubular 8, **9**
Collector spacing 4, *35*
Composition of dust, effect of 14–18

Composition of gas, effect of 11–14,
    39–51
    Components 34, 220, 222, 241
    Temperature 40, **43**
    Pressure 43, **43**
    Volume 44, **45**
    Velocity 44, **45**, **47**
Computational fluid dynamics
    gas distribution 13, 51
    electrode type and location 71
Conditioning of flue gases and dusts
    sulphur trioxide 60, 224, 228, 240
    ammonia 226, 230
    dual cond. $SO_3 + NH_3$ 60, 226, 232
    moisture/humidity 232
    hot side sodium 225
    reducing operational temperature 57, 233,
        234
Contact time 19, 44
Corona discharge
    positive/negative energisation 22–3
    breakdown voltage 22–3
    corona power 22, 63
    corona suppression 55, 70, 215
    coaxial electrode system 22, 67
    current density 66
    initiation/onset voltage 22, 39, 63, 66
    parallel plate system 66
    quasi-empirical relationships 66–8
    radius of curvature of DEs 68, **68**
    space charge effects 27, **70**, 71
    streamers 235
    threshold voltage 22–3
    see also Particle charging mechanisms
Corona wind 5
Corrosion monitoring and its impact on
        performance 239–48
Cunningham correction factor 29, **29**

Design considerations
    mechanical aspects 63–86

gas characteristics 11–14, 39–60
dust characteristics 14–19, 215–36
Detarring 8–9
Deutsch Equation 33–4, 39, 45, 47
Dew point temperature. 2, 8, 57, 220, **221**, **240**
Diffusers/screens 48, **49**
Diffusion charging 4
Discharge electrode spacing 71, **72**
Discharge electrode support 70, **77**, **78**
Dry precipitator applications 6
Duct spacing *35*, 71
Dust Concentration 14
Dust composition 15, 215
Dust Deposition Internals **81**, 82, **83**
Dust Removal
  Internals 5, 32, 80
  Hoppers 5, 85

Effective migration velocity
Electron attachment
Electrical clearances
  internal/external 79, *80*
Electrical energisation
  Mains frequency energisation 91–145, 2, 119 *see* Rectifiers, Control means, HV Transformer and Automatic Voltage Control
  Hybrid ESP arrangement 149–51
  Intermittent energisation 18, 151–6
  waveforms **153**
  operational characteristics
    voltage levels 154, *155*
    current levels 155, *155*
    collection efficiency 156–9
    normal resistance dust 156, **157**
    higher resistance dust **NO RI** 157, **158**
    severe RI operation 157, **158**
  Pulse charging methods 159–70
    waveforms **160**, **163**
    with pulse transformer 161, **161**
    without pulse transformer 162, **162**
    multipulse operation 162, **163**
    operational characteristics
      current control 166
      collector current distribution 166
      electric field strength 167
      particle charging regime 167
      power consumption 168
      collection efficiency data 169, **170**
      applications 170–1
  Two stage precipitation
    air cleaning 147
    hybrid approach 148–9
  High frequency power conversion 2
    switch mode power supplies 176, **177**
    inlet stage topologies 179–84
    high frequency inverter stage 184–8
    HT transformer design 188–90
    output rectification 190

short duration pulse operation 190, 202–3
    EPSRC design. 192, **193**
    design Supplier No 1. 194–6, **197**
      field data 195, *195*
    design Supplier No 2. 196–8, **199**
      field test data 198–203
    design Supplier No 3. 204–9, **208**
      hard switching 205
    design Supplier No 4. 209–10, **210**
      field test data 210, *211*
Electric field distribution
  plate precipitators 22
  tube precipitators 22, 64
  CFD approach 64
Electrodes
  controlled emission type 18, 69
  discharge elements **65**
  high emission types 18, **70**, *70*, 215
  mounting of electrodes 68, 76
Electromagnetic interference (EMI in HFDC) 178, 180, 182
Electropositive/negative gases 26, 39–40
Extended Deutsch equation
  See Modified Deutsch formula
Emission – Current EU legislation 1

Fabric filters retrofitting ESP casings 15
Fault finding and identification 248–64
  from electric panel lights *250*
  meter readings *251*
  test lamp operation 252
  V/I curves
    deposits on internals 220, 253, **254**, **258**
    Static/operational curves 253, **256**
    High resistivity dust characteristics **257**
    H.T. insulator, hopper build up etc. 260
    Gas distribution/air inleakage 261
    Particle size changes and opacity 262
    Systematic fault finding procedure **262–3**
Flares (gas distribution) 48
Flashover 113, 115, 175
Flue gas conditioning/systems 18, 228–33
  SO_3 228–9, *230*
  NH_3 226, 230
  Dual system 232, **232**
  Results of conditioning *232*
  Humidity 232, 223
Fly ash composition, importance of 215

Gas distribution 48–54
  CFD modelling 13, 51–4
  field testing 13
  flow bypass 49
  gas distribution methods 13
  impact on performance 45
  scale modelling 13, 48, 50
  skewed distribution 13, 49
    see also Aerodynamic factors affecting performance

Hard switching HFDC supplies 205
Heavy metals 55
High resistivity particles, *see* Particle
    resistivity
High temperature/high pressure precipitation
    13, **43**, 44
High voltage measurement
    mean kV resistance 114
    peak kV capacitance 114
High voltage supplies, *see* Electrical
    energisation
History of precipitation
    early designs and applications 89
Hoppers
    air inleakage 86
    overfilling effects of 85–6
Hot side ESPs 225
Hydrofiners 12

Instrumentation for AVC
    secondary 130
    primary 131
    opacity/energy saving 132, **134**
    energy management 133, **135**
Insulators
    HT. lead through 76, *77*
    heating & purging 259
    rapping drives 77
Intermittent energisation *see* Electrical
    energisation
Ionisation 4, 22, 25–6
Ionisation coefficient 24, **25**
Ion production 63

Laplace equation 19, 22
Linear inductors 109, 115, 120, 126, 128, **128**
Line control resistors 108, **109**
Low resistance particles 236

Materials of construction 12
    for high temperature applications 41
    for wet precipitators 41
Maxwell equation 64–5, 121
Mechanical design features
    DE formats 18, **65**, 69–70, **70**, 215
    DE separation 71, **72**
    DE radius of curvature 68, **68**
    Electrical clearances 79, *80*
    Electrode support insulators 70, *77, 78*
    Gas distribution 48–54
    Hoppers, overfilling/inleakage 85–6
    Rapping 4, 32, 48, 80
    Re-entrainment 32, 44, 48, 84, 86, 262
    TR sectionalisation 5, 75, **75**, 76
Mist precipitators
    applications and design 8, **9**
Modelling of precipitators
Modified Deutsch Formulae
    Matts Ohnfeldt 34–5, 44, **45**
    Petersen(FLS) 34

Navier Stokes equations 50

Particle charging mechanisms 4
    Cochet's model 27–8
    Impact/collision charging 4, 27–8
    ion diffusion 27–8
    saturation charge 27–8, **28**
    particle concentration 14
Particle migration 29
    exponential law 33
    practical consideration 33
    theoretical 30
Particle re-entrainment 48
Particle resistivity 1, 216
    critical value 15, 224
    effect on performance 41, 220
    effect of temperature 41, **42**, 218, **218**, **219**,
        234
    effect of sodium in ash 221, **222**
    effect of sulphur + sodium on resistivity **222**
    effect of sulphur + sodium on perform. **222**
    measurement laboratory/site 217, **218**, **219**
    Bickelhaupt prediction of resistivity 222
    resistivity change conditioning 41, 224–5,
        234
    low resistivity dust 16, 226
    surface properties of particulates 18, 216
    matrix resistivity 18, 216
    difficult dusts 19
Particle size
    distributions 16, 215
    grade efficiency relations **16**, 31
    shape 17, 216
Particle transport 30
    see Effective migration velocity
Pilot precipitators
Plasma region 27
Poisson's equation 19, 22, 123
Positive/negative energisation **23**, 26–7
Precipitator Applications 10–11, 170–1
Pulse charging *see* Electrical energisation

Rapping 4
    collectors 32, 80
    discharge elements 32
Rapping optimization 4, 32, 80, **82**
Rapping intensity and frequency 84
Re-entrainment effects 32, 44, 48, 84, 86, 236,
    262
Rectifiers
    Mechanical switch 92, **95**, **96**
    Valve cold/hot cathode 91, **92**, **93**
    Metal oxide copper oxide 97, **97**
    Metal oxide selenium 97, **98**
    Silicon 2, 98–101, **98**
Rectifier control methods
    early transformer tap changing 107
    autotransformer motorised 108, **109**
    transductor/magnetic saturable reactors
        109, **110**

thyristors 2, 110, **111**, 121
see also Automatic voltage control
Residence time 44
Resistivity *see* Particle resistivity
Reverse ionisation 15, *see* Particle resistivity
Reynolds number 29, 51
Ruhmkorff coil 89, **90**, 91
Ruhmkorff output waveforms **90**

Saturation charge on particle 27
Scouring of dust from internals 48
Sectionalization 5, 75, **75**, 76
Selective dust separation
Single stage precipitation 2–3
Sneakage and sweepage 49
Sodium depletion 225
Space charge effects 7
Space charge region 26
Sparking 138, **138**, 139
Spray irrigation 6, 32
Specific collecting area (SCA) 35, 44, **45**
Specific power input 73, **74**, **76**, 142
Stokes-Cunningham correction 29, 55
Switch mode power supplies 175–212
(*see also* HFHV operation)
Sulphur reducing bacteria 41
Supervisory AVCs 143–4

Temperature effect on performance 40–1
Temperature effect on design 12
Theory of precipitation 1
Theoretical migration velocity 30, **31**, 34
Treatment time 19, 44
Trichel pulses 27
Transformer – HV mains design
  Winding arrangements 101
  Losses 102
  Cooling of units 105

Equivalent circuit **105**
Oil testing *106, 107*
Current form factor 121
TR ratings 123
Secondary current 123–4
Primary current 125
Secondary voltage
Apparent power 125
Transformer – HFHV design
  Parasitic capacitance 188
  Magnetic leakage flux 188
  Insulation and stress management 188
  Corona effects 190
Tube type precipitators 8, 60
Turbulence in ESPs 46, **47**
Two stage precipitation 2, 3, 21, 147, **148**, 149

Ultra-violet region 27
Upgrading of precipitator performance 35, 36

van der Waal forces 4, 32, 58
Velocity of gas 44, **45**, **47**
Viscosity and density of gas 44
Volatile materials 56
Voltage and current waveforms **102**, **120**, **122**, **127**
Voltage doubler circuit **95**

Wet precipitators 6, **7**
  applications 57–8
  collector film flow 32
  spray irrigation 6, 32
  washdown system 6, 32
  water treatment 57
Wimshurst & Voss generators 21

Zero crossing (thyristor switching) 119